GALLERY OF SCHOLARS

Philosophy and Education

VOLUME 13

SCOPE OF THE SERIES

There are many issues in education that are highly philosophical in character. Among these issues are the nature of human cognition; the types of warrant for human beliefs; the moral and epistemological foundations of educational research; the role of education in developing effective citizens; and the nature of a just society in relation to the educational practices and policies required to foster it. Indeed, it is difficult to imagine any issue in education that lacks a philosophical dimension.

The sine qua non of the volumes in the series is the identification of the expressly philosophical dimensions of problems in education coupled with an expressly philosophical approach to them. Within this boundary, the topics—as well as the audiences for which they are intended—vary over a broad range, from volumes of primary interest to philosophers to others of interest to a more general audience of scholars and students of education.

The titles published in this series are listed at the end of this volume.

Gallery of Scholars

A Philosopher's Recollections

by

ISRAEL SCHEFFLER

KLUWER ACADEMIC PUBLISHERS
DORDRECHT / BOSTON / LONDON

A C.I.P. Catalogue record for this book is available from the Library of Congress.

ISBN 1-4020-2709-5 (PB)
ISBN 1-4020-2679-X (HB)
ISBN 1-4020-2710-9 (e-book)

Published by Kluwer Academic Publishers,
P.O. Box 17, 3300 AA Dordrecht, The Netherlands.

Sold and distributed in North, Central and South America
by Kluwer Academic Publishers,
101 Philip Drive, Norwell, MA 02061, U.S.A.

In all other countries, sold and distributed
by Kluwer Academic Publishers,
P.O. Box 322, 3300 AH Dordrecht, The Netherlands.

Printed on acid-free paper

Printed in the Netherlands.

In memory of

Sidney Morgenbesser

Friend Through the Years

TABLE OF CONTENTS

PREFACE

This book offers a personal account of scholars in philosophy and education with whom I have had the good fortune to interact during the course of my half century at Harvard University and elsewhere. My aim in writing this account is threefold: first, to recapture for myself the pleasure of their memorable company for its own sake, secondly, to have occasion to reflect on the educational impact of their teaching, and, finally, to counteract the prevalent amnesia of universities by recalling the conduct of scholars of past generations who still have things to teach us.

I owe thanks to many people who have helped me in this endeavor. Professor Harvey Siegel, Dr. Stefania Jha, and Dr. Rosalind Scheffler read initial versions of the manuscript and gave me the benefit of their criticisms, as did the publisher's anonymous readers. JoAnne Sorabella and Stefania Jha listened to my readings of a number of these chapters, and JoAnne Sorabella produced several typescripts of the whole with her usual matchless proficiency. I presented some portions of the manuscript to the Philosophy of Education Research Center at Harvard and profited from these occasions. After I joined the Mandel Center at Brandeis University in the Fall of 2003, Avital Feuer assisted me ably in readying the final version of the book. And I am grateful to Laurie Scheffler for her meticulous help with proofreading.

A number of people gave me invaluable aid in connection with the album of photographs included here. Robin McElheny and Andrea Goldstein of the Harvard University Archives; John Collins III and Marylene Altieri of Harvard's Gutman Library; Abby Lester of the Columbia University Archives—Columbiana Library; Sharon Lehner of the New York University Archives, Photographic Collection; Linda Jack, Kathleen Much and Jenny Hayes of the Center for Advanced Study in the Behavioral Sciences at Stanford; Claudia Kishler of the Ansel Adams Publishing Rights Trust; Helen Killick of London; Patricia White and Judith Morrison, of the Institute of Education at the University of London;—all provided me with indispensable assistance in securing the photographs I needed. Particular acknowledgement for each photograph is given on the page on which it appears.

Finally, I express my profound thanks to the scholars themselves, whose work and style provide lasting lessons for further generations to ponder.

Israel Scheffler
Brandeis University
Spring, 2004

CHAPTER 1

INTRODUCTION

Research universities have no memory. They properly educate the young in the heritage of the past in order to equip them to criticize, revise or expand it in the future. Research in every theoretical discipline looks forward, wasting no time in dwelling on bygones, especially bygones concerning its own activities. Each student cohort sees only that time slice of university culture coordinate with its own residence in academia, affecting the conditions of its own stay. Each student, moreover, is engaged in expanding personal knowledge, enhancing relevant skills, and planning a career, every such task oriented toward the future. As a result, the past cultures of research universities are typically lost from view, along with the personalities, attitudes, and ideas that shaped them. Only those whose time perspective antecedes that of the student cohort have the capacity to recall portions of the university's past as embodied in their own experience.

My motivation in writing this memoir is to recapture some of my academic past, from my first appointment as a Harvard instructor in 1952 until the closing of my philosophy of education research center at Harvard in 2003. As a member of the faculty of education charged with teaching philosophy of education, I have been especially concerned with the ideas embodied in teaching practice and with the attitudes of those responsible for carrying it forward. Such ideas and attitudes are not fully expressed in the catalogues issued by university schools or departments, nor are they fully explicit in the statements of professors and administrators themselves. These ideas and attitudes are often revealed obliquely, exemplified by what such authorities say and do as they go about their normal business. And such exemplification has to be read in context, a product of the subtle play of personal factors within a particular academic environment.

"Man is a symbol", Peirce has somewhere said; the ideas and attitudes a person symbolizes reveal themselves only to interpreters cognizant of their own limits and aware as well of possible alternatives. That there are indeed limits as well as alternatives is far from saying that every interpretation is unwarranted. On the contrary, not every interpretation violates proper limits, nor is the warrant of any interpretation lessened by the existence of appropriate alternatives. At any rate, I intend here to provide personal accounts of some of the memorable university personalities it has been my good fortune to interact with during a long academic journey. My aim is, first, to retrieve them from memory for their own intrinsic interest and, secondly, to highlight aspects of their conduct that I believe exemplify important values, and educational ideas or attitudes worth recalling for new generations to ponder. Inevitably, in trying to recall these personalities, I will need to attempt a description of the context within which they flourished.

In an earlier book, *Teachers of My Youth*,[1] I offered accounts of my early educational experience, with special reference to my Jewish education in family and school. That book also presented portraits from memory and my caution there is applicable here as well. Such portraits, I said, "have the defect that they cannot be rechecked by others directly, although my memories can in principle be strengthened or weakened by indirect means, e.g. the memories of others, the records of the schools involved, general information about the context etc. I have nevertheless not tried to test my memories in this way and am aware that what I deem recollection might in fact turn out to be inaccurate in one or another respect. Not deterred by this possibility alone, I thought it worthwhile to put down whatever I seem to remember, with the caution that it is surely open to revision or withdrawal should contrary evidence turn up. My putative memories are, at any rate, not worthless just because they are in principle refutable."

In *Teachers of My Youth*, I noted two books that encouraged me in my project, *Great Teachers*, edited by Houston Peterson,[2] and *Masters*, edited by Joseph Epstein.[3] Both are composed of chapters by former students about their teachers and I found both to be enormously interesting. "Such biographical essays, essentially from memory," I wrote, "seem to me to have seen more deeply into aspects of education than many other, generalizing forms of investigation. Analogous forms of narrative description are available, to be sure, both in history and in fiction, but they differ in one essential respect: neither history nor fiction need be based on memory."[4]

Despite the similarity of approach between that book and the present one, there are critical differences. That book concerned my early youth; it was written at a much greater distance from the events and experiences recounted. Moreover, it dealt with the culture of Jewish immigrant generations from Eastern Europe and had as a main purpose to bridge the cultural divide separating that culture and its religious and educational outlook from my adult experiences and perspective. The present book concerns a period much nearer the present. It begins with my graduate schooling and extends through my time at Harvard. Its focus is not cultural but rather individual; it attempts to capture the educational personalities of those who exercised a profound influence on me during this period, and in so doing to provide a sense of their academic environments.

As in the previous book, although much of what I write here is bound to be autobiographically revealing, I do not aim to produce an autobiography. My attention, as before, is directed outward, toward the teachers and colleagues I describe and their forms of teaching and scholarship. I hope that this outward focus will help to keep my story as accurate as possible. Anything that touches on oneself is bound to be subject to distortion, defensiveness, and wishful thinking. I have however tried to tell the truth and hope that even where my memories fall short, they may at least bring out points of educational interest useful in their own right.

My earlier book ended with my undergraduate years, my stated reason being that to have carried my treatment further would have changed my memoir into something else, involving a large admixture of my current attitudes toward contemporaries and philosophical issues. In the present book, I have decided to override my previous compunction. Concerned less with cultural matters than with particular persons and with a time frame closer to the present, this book has less the character of a memoir disclosing my own early formation than the character of an album of portraits of memorable scholars with whom I interacted as an adult. In any case, I am prepared now to accept the consequence that my treatments will reveal some of my own current attitudes toward educational and philosophical matters.

CHAPTER 2

SOME NEW YORK PHILOSOPHERS

The first graduate philosophy courses I had were taken in the late forties at Columbia with Professor Ernest Nagel and dealt with logical theory and philosophy of science. I had been predisposed toward these topics by my earlier studies with, and readings in analytically inclined philosophers and my initial admiration for scientific forms of reasoning. In my undergraduate days, I had been influenced by the teaching of Laurence J. Lafleur, a rationalist philosopher and commentator on Descartes, who wrote on logic and several scientific fields. I had also had courses with the great historian of science, Alexandre Koyré, on the rise of modern science. In my independent reading, I had studied, among other works, Russell's *The Scientific Outlook,*[1] A. S. Eddington's *The Nature of the Physical World,*[2] H. Dingle's *Through Science to Philosophy,*[3] H. Reichenbach's *From Copernicus to Einstein,*[4] Susan Stebbing's *Philosophy and the Physicists,*[5] Whitehead's *Science and the Modern World,*[6] Tarski's *Introduction to Logic,*[7] M. R. Cohen's *A Preface to Logic,*[8] A. E. Murphy's *The Uses of Reason,*[9] S. Hook's *Reason, Social Myths and Democracy,*[10] and Cohen and Nagel's *Introduction to Logic and Scientific Method.*[11]

The latter three authors were well known and much admired in New York City where they taught, both Hook and Nagel having been early students of Cohen at the City College, with Hook later teaching at New York University and Nagel later teaching at Columbia. Both Hook and Nagel came under John Dewey's influence in the later course of their careers, Hook remaining Dewey's staunch advocate and defender, earning the sardonic sobriquet "Dewey's bulldog", and Nagel, a sympathetic critic of Dewey's *Logic,*[12] being awarded the Columbia title, "John Dewey Professor of Philosophy".

Both Dewey and Cohen were proponents of scientific method in philosophy, and transmitted this emphasis to their students. Both Dewey and Cohen had, in their turn, inherited their scientific bent largely from C. S. Peirce, the founder of American pragmatism. Dewey had, in fact, studied for a time with Peirce at Johns Hopkins where, he later testified, he had not gotten much from such study at the time but, many years later, saw the point of Peirce's teaching and eventually developed his notion of "warranted assertibility" in the spirit of Peirce, as he understood him. Cohen, for his part, had written his thesis at Harvard on Kant, the main influence on pragmatism, and was heavily influenced by Peirce, later producing the first anthology of Peirce's papers under the title, *Chance, Love and Logic* (rather than his originally proposed title, "Tychism, Synechism and Agapasm").[13]

But Peirce, a brilliant polymath who struck original sparks in over a dozen fields of learning, presented different messages to different readers. Both the architect of pragmatism and a traditional realist, he interpreted science, in light of his pragmatic maxim, as clarifying ideas by reference to actions and consequences, but at the same time held science to be a community of investigators advancing inevitably toward an ultimate agreement definitive of truth, its object a reality independent of what any particular person or group may think. To oversimplify matters, Dewey inherited the pragmatic Peirce, Cohen the realistic Peirce. While both enthusiastically defended scientific method, Dewey treated such method as an instrument for resolving the problems set to an organism or a group of organisms by environing conditions while Cohen treated science as seeking universal laws through rational means.

Dewey thus took his cue from psychology, biology and history, treating logic itself as an evolutionary product for resolving the difficulties of organisms, while Cohen took his inspiration from mathematics as representing universals, within which empirical experience was to be organized in networks of natural laws. Symbolic of the contrast is a comparison of the two titles chosen by Dewey and Cohen for major works of theirs. Dewey called one of his important books *Experience and Nature*,[14] whereas Cohen called his book *Reason and Nature*.[15] While for neither thinker is nature an initial given, for Dewey it is to be approached through experience, whereas for Cohen the road to nature is reason. Dewey justified logic and scientific method as evolutionary emergents of paramount use to organisms in overcoming obstacles to their survival and growth. Cohen criticized such justification as an acknowledgement of the value rather than the dignity of thought.

A word must be addressed to the general culture within which all our protagonists worked. It was, in brief, a broad intellectual culture. The common emphasis on scientific method which all shared by no means portended a technical specialism. The idea was that a common thread of responsibility to both the canons of logic and the demands of empirical evidence unifies all the branches of learning, science, history, arts, politics, philosophy and the law. This breadth of interest in the many departments of civilization was shared by Dewey, by Cohen, by Hook and Nagel, and it was supported by the university spirit of both the City College and Columbia, which encouraged breadth of general learning and concern for the topical issues of contemporary society. All the thinkers we have been concerned with flourished under this emphasis, while embodying it in their variant ways.

Dewey propounded a large vision of philosophy as a critical element of culture, whose historical task it is to respond to the major problems and opportunities of each age in which it is situated. It is the problems of men (the title of one of his collections), rather than the problems of philosophy to which its energies are properly to be addressed. A constructive thinker, he developed original views of thought, knowledge and value and tested them by addressing urgent issues of education, democracy, and social reconstruction. Cohen was enormously learned but philosophized primarily as a critic whose work as reflected, for example, in his *Reason and Nature* and *The*

Meaning of Human History,[16] grew out of an acute dialectical appraisal of doctrines already formulated.

The teaching styles of Dewey and Cohen were characteristically different. Dewey tended to think while lecturing, to develop the thread of an argument in front of his class, knowing where he was ultimately headed but working out the details in process. His lectures on philosophy of education in 1899 have been preserved in a student transcription, published later as edited by Reginald Archambault. These lectures, precursors of his finished publication, *Democracy and Education*,[17] show his teaching style in action. Even where a given lecture ends, seemingly conclusively, the next hour picks up the thread and carries it onward, through various ups and downs, in a direction initially unforeseen but autonomous and determining the architecture of the whole.

Cohen's style, as attested by various of his students, was quite different. He did not propound a large positive structure but offered penetrating analyses of extant views. Unlike Dewey's style, which interacted little with his auditors but ran along its own track, Cohen's style was abrasive and combative, even causing him to be thought needlessly negative toward students in class. That he was witty and acute and erudite on a surprising number of topics was well known and earned him the esteem and admiration of the scholarly community. That his tough methods of teaching trained a large variety of students in severely critical attitudes that helped them to academic heights cannot be gainsaid.

Perhaps the flavor of some of the apocryphal stories that grew up about him in New York will give a sense of his teaching persona. One such story relayed to me by my friend Sidney Morgenbesser concerns a lecture he was purported to have given, entitled "Twenty Theories of Crime", during the height of the depression of the 30's. To provoke his students, many of whom were radicals of one or another sort, he omitted the one theory they were most interested in, the economic theory. Having concluded his survey and critique of all the other theories on his list, Cohen waited for questions from the audience. City College students of the period, not reputed for deference to their elders, began the questioning in the combative spirit for which they were well known: "Professor Cohen, have you read a book in sociology in the last twenty years?" Reply by Cohen: "Has there been a book in sociology in the last twenty years?" Next question: "Professor Cohen, has it ever occurred to you that a large number of the people in prisons are poor?" Answer: "Has it ever occurred to you that a large number of people outside of prisons are poor?"

Cohen's erudition was legendary and no doubt provoked the envy of students and peers alike, who also felt the sharp edge of his critical remarks and resented what they considered his one-upmanship. This story, which I heard from a philosophy instructor when I was still an undergraduate, purports to tell of a monthly luncheon group of faculty colleagues at the City College. Cohen's colleagues were, according to the story, constantly and increasingly annoyed by the fact that, no matter what topic came up for discussion at the meetings of the group, Cohen always had the last word. His colleagues finally conceived of a plan to put him in his place. They decided to pick some abstruse topic in advance of a forthcoming meeting, and to prepare their respective comments in line with their own specializations. The topic they fixed on was ancient Chinese pottery and they proceeded accordingly to prepare their comments for the next meeting.

When the time came, the first comment, over coffee, came from a professor of chemistry, who lectured the group on the chemical composition of the porcelain in ancient Chinese pottery. Thereafter, the professor of art delivered a talk on the styles represented in such pottery, the linguistic professor discussed the linguistic character of the inscriptions on ancient Chinese pottery, and the professor of archaeology spoke on the archaeological evidence for the dating of such pottery. So they continued, one after another, to discourse on the chosen topic from their respective special expertise and, to their surprise and increasing glee, noticed that Cohen had not said a word, neither interrupting the speakers nor adding any comments of his own. Finally, not able to contain himself any longer, one faculty colleague turned to Cohen and said, "Professor Cohen, have you nothing to say about our topic today?" Whereupon Cohen replied, "I'm so pleased, gentlemen, that you have all read my article in the Encyclopedia Britannica on ancient Chinese pottery, but I have changed my opinions since I wrote it."

Both Hook and Nagel were students of Cohen and profited from his teaching although Hook later came to criticize both Cohen's teaching style and his realism, becoming a life-long advocate of Deweyan pragmatism. Nagel, for his part, departed from Cohen's style, as well as from his views on logic, under the influence of the new formal and positivistic trends that flourished with the arrival in the U.S. of such eminent scholars as Carnap, Tarski, and Hempel, who had managed to escape from Europe before the war years. Hook worked primarily in post-Hegelian history of philosophy, social theory, and philosophy of history and education, becoming well known both as an interpreter of Marx and a trenchant opponent of Stalinism and Soviet totalitarianism. Nagel, by contrast, worked in the areas of logic and philosophy of science, theory of measurement and probability, as well as history of science, offering a blend of classical views on these topics with pragmatic interpretations.

The ideal of erudition exemplified by Cohen was evident in both Hook and Nagel, who impressed on their own students the value of broad humane learning along with the indispensable art of logical appraisal and evidential accuracy. Nagel had written widely on topics in the history of mathematics and science, had composed an important monograph on the theory of probability, and was a prolific reviewer of philosophical books. In one lecture course I had with him, he lectured on the imageless thought controversy in psychology, on issues in logical theory and on explanatory models in economics. In collections of his papers, he included, among other topics, a critique of Brand Blanshard's idealism, an assessment of Peirce's views on induction, and a defense of logic without ontology. When, in 1961, his major book,

The Structure of Science,[18] appeared, it was plainly evident how wide-ranging his interests were, inclusive of the natural and the social sciences, psychology, biology, and history.

Nagel's lecture style was systematic and lucid. Clarity was his guide. He had neither the autonomous system-building tendency of Dewey nor the combative thrust of Cohen. He did not present a dramatic approach to his subject but captured his audience by the straightforward and transparent honesty of his presentation. It was not for nothing that one commentator referred to him as Columbia's gentle logician.

In exchanges with students, his manner was uniformly polite and respectful of the questioner's point of view. His critical analysis of students' questions was always penetrating but never personal. Moreover, he did not begin his critical response to a question until he had made quite clear what sense the question might plausibly be interpreted to have. This care he took with student queries and comments was of a piece with his respect both for his interlocutor and for the logical bearing of the points being raised. The gentleness and the critique in his demeanor were thus two sides of the same coin.

The educational model he presented to us emphasized the primacy of such care–sympathetic interpretation of the alternative meanings a question might have, coupled with sharp analysis of the variant logical issues revealed by such interpretation. The price of this valuable educational emphasis was, however, the patience required to live by it, at least under Columbia rules in those days. The practice, as I remember it, was to allow unrestricted auditing. This meant that no one was required to register for a course in order to be admitted to regular lectures. As a consequence, in a sizeable room such as the one assigned to my class with Nagel, many of the seats were occupied by students not officially taking the course. I conjecture moreover that, on an average cold winter's day, a number of passersby on Broadway must have discovered that one could rest one's weary feet for a spell by sitting in on a lecture course in Philosophy Hall.

The result was that, in addition to pointed and relevant questions addressed to our instructor by those of us who were well prepared, various ill-motivated or otherwise irrelevant questions came at him from left field. Instead of cutting off such questions peremptorily, Nagel would typically listen to them politely and proceed to disentangle the various germane senses one could, by a stretch of the imagination aided by a liberal application of the principle of charity, assign to the original queries. Only thereafter would he proceed to respond to these queries under their variant guises and provide penetrating analyses of each. In the meantime, an almost audible groan would arise from the throats of the registered students, eager to hear Nagel's systematic development of his topic without digressions into irrelevancies, instructive as his careful disposition of such irrelevancies in fact was.

The moral of the story is that educational styles have their advantages and disadvantages; each such style has its price. Cohen, with whom Nagel had served a long apprenticeship during their co-authorship of *Introduction to Logic and Scientific*

Method,[19] represented both a thoroughgoing commitment to critical analysis and an aggressive, argumentative style of teaching. Nagel retained the critical analytic commitment but relinquished the abrasive teaching style. Cohen, who did not suffer fools gladly, ran the risk of embarrassing or even humiliating innocent but thin-skinned students; Nagel, who responded civilly to both the wise and the foolish, ran the risk of giving undue time to irrelevancies. My guess is that Cohen's risk was the harder to overcome, being a likely concomitant of his personality, whereas Nagel's risk might have been mitigated by stronger rules limiting access to courses to those officially registered and presumably well prepared. Such limitation would, in its turn, however, have incurred the loss of Nagel's admirable example of a scholar willing to address himself not only to the initiated but also to the outsider, thus representing philosophy as a humanistic effort and not just a specialism.

That undeniable educational virtues have their downsides was evident in another, more subtle aspect of Nagel's influence on students. The virtue in question here is erudition, another quality which Nagel shared with Cohen, but which, as distinct from Cohen, he exemplified in such a sympathetic and patient way as to arouse admiration and affection, rather than argumentativeness or ambivalence. His critical character seemed to us further to guarantee that his erudite knowledge had run the gauntlet of his acute logic and thus reinforced the tendency of students to agree with his views.

In the Columbia context, the ideal was humane learning, special knowledge over a broad range, critically mastered, and appreciated not only for itself but also for its cultural and historical import. This ideal was admirably embodied, in the forties and after, by three faculty members referred to at Columbia as the three wise men, the physicist I. I. Rabi, the historian of art Meyer Schapiro, and the philosopher Ernest Nagel. Their learning was so evidently deep as well as broad, their mastery so patent, and their character so congenial as to magnify the distance between themselves and ourselves, aspiring novices in the worlds of scholarship.

The range of Nagel's erudition and his acute exposition of issues arising in one or another area of his interest made it difficult for us to imagine that there was anything worthwhile left for us beginners to do. Surely, his views, already worked out with such care, over such an enormous range of philosophical concern, left nothing over in the way of unsolved problems for our generation to tackle with any hope of success. His expository lectures were emblematic of his knowledge and dealt with topics with which he had already come to terms. Our admiration for his achievement correspondingly sapped our ambition to go beyond, to explore unknown territories.

Erudition is conservation, the treasuring of what is known, and it countervails the opposing tendency to take risks in dangerous realms posing as yet unsolved problems. Clearly both conservation, which guards the knowledge already won, and exploration, which risks unsettling such knowledge in search of the new, are both essential to learning. Without exploration, already acquired knowledge stagnates; without conservation, successful explorations evaporate the moment they are concluded. The balance of the two is as precious as it is difficult to achieve, and an overemphasis on the one is a diminution of the other. The balance is rarely attained by given individuals; it is enough if what Peirce called "the community of investigators" can approximate it. It is a fortunate educational career if it includes studying with scholars of both the strongly conserving and the strongly exploring sorts. In the latter respect, I was myself extremely lucky.

I should here append a further comment on Dewey, whose influence on American philosophy and on education was unparalleled, and whose work occupied me a good deal and culminated in my book *Four Pragmatists*.[20] I had my first taste of his writing when, as an undergraduate, I bought a copy of his *Human Nature and Conduct*[21] in a college bookstore. Taking it home and opening it to the first page, I was baffled, shut the book and left it shut for about four years. Thereafter, intrigued primarily by Hook's mention of him in the undergraduate course I had with him at the New School, I tried again. This time I persevered over a long period, finding unexpected treasure in the work, and proceeding to study a considerable number of Dewey's other writings, in all of which I found philosophical acuteness and human sensitivity.

I never met Dewey in person but did in fact hear him speak on three occasions at Columbia. The first of these was a celebration of the work of the American educator Boyd H. Bode. I remember nothing of the speeches in Bode's honor but I do have a vivid recollection of Dewey at that event. Seated along with the other speakers at the rear of the Teachers College stage during their presentations in appreciation of Bode's work, he awaited the honoree's response as Bode strode to the dais and prepared to speak. Then, a catastrophe–Bode's lengthy typescript flew off the podium and its pages were scattered across the length of the stage behind him. Dewey, then advanced in years, was the first on his hands and knees, gathering up as many pages as he could hold to hand up to Bode at the podium.

The second occasion was a speech Dewey gave at a conference of medical professionals at Columbia. Of his talk I remember only the introduction, in which he said that he had no claim to speak as an authority on medical procedures, but that he could certainly speak as having been repeatedly the object of such procedures, and as one grateful for their efficacy.

The last occasion was a speech Dewey gave at Columbia in the last year of his life. If the first occasion I cited showed his modesty and the second his appreciation, this third was an illustration of his boldness and vision. The topic he chose was something like "Challenges to Democracy in the Next Hundred Years", and the two challenges he discussed were those of totalitarianism and clericalism which, he warned, would continue to pose mortal dangers to democracy long after the end of his lifetime.

About ten or fifteen years after I last heard Dewey speak, I was invited to give a talk at a Columbia department of philosophy seminar. In my earlier experience as a student of Nagel, I had often attended such seminars on Thursday afternoons, featuring presentations by visiting philosophers. The seminar room had a large central table. The speaker sat at the head and the auditors were seated around the table in two circles.

When, after a lapse of so many years, I now entered the room as speaker, I was suddenly struck by an item I had not remembered. On the wall at the back of the room, behind the speaker's chair, was a large black-and-white photograph of Dewey. When I sat down and prepared to read my paper–a critique of Popper's philosophy of science—I turned my head and saw Dewey peering over my shoulder as I began to speak, in a style quite different from his own if not altogether alien to his spirit. I was conscious of his presence throughout my talk and couldn't help wondering how he might have taken my remarks that day.

CHAPTER 3

NELSON GOODMAN AT PENN

When my graduate studies at Columbia shifted in the early fifties to the University of Pennsylvania, I became a student of Nelson Goodman, who, in the educational matters I have been discussing, was virtually the mirror image of Nagel. They of course knew one another and had interacted a bit at the New York Philosophical Circle, where Goodman had first delivered his brilliant paper, "The Problem of Counterfactual Conditionals".[1] Both philosophers were ardent advocates of logical and scientific methods and opposed to mystical and existential trends which were gathering strength after the end of the Second World War. Nagel is reported to have written one of the strongly positive letters recommending Goodman for his appointment at The University of Pennsylvania in the forties, and Goodman acknowledged Nagel in the foreword to his first major book, *The Structure of Appearance*,[2] which appeared in 1951. Despite their commonalities, however, their educational impacts were diametric opposites and the contrast struck me at my first contact with Goodman.

Goodman's reputation as an astringent philosopher had appealed to me even before my first contact with him. I had heard about him from Sidney Morgenbesser who had had some courses with him at Penn and was enormously impressed by his logical ingenuity and critical acumen. I had also read his brief papers, "A Query on Confirmation"[3] and "On Infirmities of Confirmation Theory",[4] which seemed to require a radical rethinking of scientific method. Having grown up with Deweyan pragmatism and the philosophy of science as taught by Ernest Nagel and Sidney Hook, and influenced at a distance by their teacher, Peirce's first anthologist, Morris R. Cohen, I was amply predisposed to a no-nonsense approach to phiosophy, to a rejection of absolutes and certainties in favor of scientific skepticism and the rule of logic. I was thus elated at the prospect of studying with Goodman.

I had signed up for his graduate seminar, the title of which I have forgotten but the focus of which was induction. Eager to begin my studies with him, I got to the assigned seminar room early, and was thrilled to find only about twenty-five people in the room, seated around a large table. How wonderful, I thought, in contrast to the large numbers in the Columbia classes I had experienced. Here, I could surely expect more serious interaction with the professor, not to mention my fellow students.

A few minutes after the beginning of the hour, Goodman entered, looked around the room at the assembled students and said, "This group is too large. After the break, half of you will be gone." He then proceeded to give us a long list of recent journal articles on induction and announced that we would be expected to make individual reports to the seminar on articles we were to choose. After some general remarks on the seminar theme and some further explanations of the procedure, Goodman took questions from the group and announced a mid-session break. And, lo and behold, when the seminar resumed, the group had indeed dwindled to about fifteen, including, among others, John Fisher, Sidney Axinn, Samuel Shuman, Beverly Robbins, Noam Chomsky and Elizabeth Flower.

In further contrast with my Columbia classes, Goodman's seminar did not consist of lectures by him. The onus was on us to respond to the chosen articles and to engage in the ensuing deliberations about the substantive points raised. Nor was Goodman perceived as particularly sympathetic. He was certainly not unpleasant in any way, but he was all business and his business was philosophy. Moreover, we soon came to understand that philosophy here meant the solving of philosophical problems, not metaphilosophy, or history of philosophy, or applications of philosophy to society.

This single-mindedness had the effect of intimidating many students and would-be students who had heard of Goodman's reputation, a reputation that tended to inspire not affection but apprehension. Concomitant with his single-mindedness was Goodman's evident impatience. In one of our first seminar sessions, a student assigned to one of the listed articles proceeded to report on its first three sections, pointedly omitting its fourth and final section dealing with mathematical aspects. When she had concluded, Goodman turned to her and asked, "Why didn't you report on the last section of the article?" The student replied, "Because I don't have the background", to which he immediately rejoined, "How long would it take you to get the background?" Rather than proceeding to acquire the background, that student left the class after the session was over, never to return.

The most profound effect of Goodman's style had to do with the problems he dealt with. These were not problems he had already solved to his satisfaction, given to us as mere pedagogical exercises, or else as illustrations of large themes he was prepared to expound. On the contrary, he gave us problems with which he was himself still actively engaged, and to which he had as yet found no solution. The double lesson his style conveyed was, first, that there were still unsolved problems in philosophy, to which even the most renowned thinkers had as yet found no answers, and second, that we novices could participate, along with such thinkers, in the quest for solutions. The distance between such thinkers and ourselves, rather than being magnified, was reduced to the vanishing point. Rather than being awed by their achievements with the numbing effect of losing our will to go forward, we were inspired by their evident ignorance to explore for ourselves the perilous terrain of open problems. It was no accident that several publications emerged from student papers in that seminar of Goodman's that I have described, including Chomsky's first paper on syntax and my two first papers in philosophy.

I have painted the contrast between Nagel and Goodman in perhaps too harsh a light. Nagel certainly contributed important solutions to central problems, e.g. interpretations of the theory of probability, analyses of functional and historical explanation; and Goodman saw his own solutions as continuing a dominant trend from Hume and

Kant through pragmatism to the theory of symbols. Moreover, solutions are, in any case, to be conserved if they are worth pursuing at all, and conserved solutions are in turn grist for further advances if they are not simply to die of inanition. Yet the operative teachings of Nagel and Goodman were so strikingly contrastive as to bring out forcefully the essential importance of both vital moments in the development of learning, the focussed problem-solving drive and the reflective appreciation of its larger context and significance.

Nor have I perhaps done proper justice to Goodman in describing his business-like and single-minded attitude in teaching. He did not, in general, enjoy lecturing. His dominant method, rather, was live interaction with students, listening to what they had to say, responding to them, raising questions, offering proposals, criticisms, and so forth. He did not offer literal, sequential accounts of the propositions comprising his subject matter, nor were his contributions dry and empty of affect. Often they were emotively sharp, sometimes pungent, typically challenging, occasionally witty, but always in the service of a common pursuit of understanding. The provocative atmosphere of discussion in effect drew students into the center of the issues as equals, leaving little room for distancing themselves from the action. Goodman's method of teaching the introductory philosophy course was no different from the graduate seminar I have described, except for the level of the students who, in both cases, were stimulated into entering the process of trying to solve problems of philosophy by their own lights.

The positivistic approach to philosophical communication seemed to me quite different. This approach, increasingly influential in the forties and fifties, was best represented in the writings of the eminent positivist philosopher Rudolf Carnap who strove always, no matter what the topic, to provide the clearest possible systematic account of his views, exposing his first principles and chains of argument in quasi-mathematical form. Something of this attitude, though less extreme, characterized also the presentations of Nagel, who had himself come under the strong influence of Carnap. The positivists made a radical distinction between cognitive and emotive meaning and propounded the view that the emotive variety had no place in science nor in any philosophy that purported to be scientific. Their preferred forms of exposition were, accordingly, as literal and emotively neutral as possible, clear and explicit but dry and, of course, humorless.

I still remember what a shock it was for me, in my first year of graduate work, to study *Methods of Logic*,[5] the textbook published in 1951 by W. V. Quine, the celebrated Harvard logician and sometime collaborator of Goodman. Having earlier studied the logic text of the mathematician Alfred Tarski and a couple of Carnap's books, I had naturally assumed that humor and logic were sworn foes, never to be caught together within the covers of a single book. Quine's text took me aback for there, almost from the beginning, humor took its regular place in making logic crystal clear to the reader. Quine's motto was evidently: clarity yes; deadpan literalism no. One could apparently have fun while learning logic, and without the slightest loss in transparency to boot.

I still recall with pleasure Quine's names for three important logical maneuvers he was expounding. Rather than giving them accurate but boringly dull tags, he called them "the fell swoop", "the full swap", and "the full sweep".

Both Quine and Goodman, in any case, departed from the positivistic practice of philosophical communication, with Goodman, in his *Languages of Art*, [6] propounding the radical view that emotions, far from being hindrances to cognition, are actually important cognitive instruments. And his own writing more and more employed metaphor, alliteration, and wit in getting his points across. That he took, not a solemn, but a somewhat ironic view of his own teaching and writing was illustrated at a publishers' party called to celebrate the Festschrift for him, *Logic and Art: Essays in Honor of Nelson Goodman*,[7] edited by Richard Rudner and myself.

After the usual pleasant introductory remarks by the publisher and others, Goodman was called upon to reply. He responded somewhat as follows,"When I received this Festschrift and saw how many eminent scholars were included, I could not imagine how the editors had managed to get this illustrious group to contribute. Who among them could have been expected to be willing to write a tribute to Nelson Goodman? Then, after puzzling over this for a while, it came to me: the editors approached Alonzo Church and asked him if he would contribute to a volume for Nelson Goodman, whereupon Church said, 'Not on your life!' The editors responded, 'But, Professor Church, W. V. Quine has already agreed to contribute'. Church then said, 'Well, in that case, count me in.' Then the editors went to Quine and invited him to write an essay for the Festschrift, whereupon Quine said 'Absolutely not!' 'But', said the editors, 'Church has already agreed', to which Quine said, 'Of course if Church is going to contribute, so will I.' So the editors continued, inviting each remaining pair of contributors who, by the same device all agreed, thus completing the book".

In further remarks on the same occasion, Goodman made mention of his classroom teaching, somewhat as follows, "I have noticed how popular it has become to ask students to evaluate their instructors and their courses at the end of the term, so I decided at the conclusion of the last term to follow suit. I constructed a set of questions and gave them to the students to answer anonymously and return to my mailbox with their replies." He then read from some of the student replies, several of them humorous. But the one he saved for last was a reply that began, "At first, I thought the instructor was a mean man . . . " whereupon Goodman paused and then continued, "It's the 'at first' that bothered me."

Despite Goodman's reputation as an intimidating instructor, he nevertheless managed to attract a small but devoted following of graduate students. Such students overcame their initial apprehension at his manner, seeing it as a natural concomitant of his seriousness in tackling the most central problems. Even his impatience had a bracing effect. Unlike the attitude of many faint-hearted philosophers and students of philosophy who thought that all central problems had already been treated in the history of the subject and all theoretical positions already spoken for, Goodman's impatience exuded optimism; it looked to new solutions for old problems and new problems to replace the old chestnuts. This optimism meant that we students might actually hope to discover something new or say something that had not yet been said.

To students who adopted Goodman's problem-solving ethos, he gave strong support. Indeed, he was unusually accommodating even to prospective students who indicated an eagerness to work with him. At least that is the only explanation I can muster for his response to me when, newly arrived in Philadelphia and eager to start my studies at Penn, I took the unusual liberty of phoning him at home, never having met him. "Professor Goodman", I said, "I've just moved to Philadelphia and enrolled as a graduate student at Penn, primarily to study with you. Now I find that your seminar meets only once weekly, from 4:00 p.m. to 6:00 p.m. and I cannot stay past 5:40; can I still be admitted?" After a short pause, he replied, "That's no problem. I usually take a break in the middle of the session, but I'd be glad to move it down to the end and conclude at 5:40."

That a student, once admitted, was hard working was a point of high praise in his lexicon of appraisal. He wrote individual comments on the student papers he required in his seminars, acute but short remarks on the major issues raised, and supportive in tone. Live reports in class were listened to carefully, and although Goodman participated in the subsequent discussion he was careful not to dominate or skew their direction, taking obvious pleasure in the initiatives displayed by other members of the seminar.

On occasion, he would note on a paper that it might be worth submitting for publication. Twice in my first seminar with him, I was elated to receive such a comment. Revising each of the papers in question in line with a couple of his written criticisms, I showed him my revisions and asked for his opinion. He surprised me by refusing. "If I were to give you another set of comments, it would be my paper not yours. It's your paper, you decide what you want it to contain; when it's published, you will be able to consider it yours."

He followed a similar policy at the later stage of thesis advising. He told me at one point that I ought to decide on a thesis problem. We had no further discussion of the matter. When I had fixed on a problem, I told him that it concerned the interpretation of indirect discourse. He approved of my idea, whereupon I proceeded to prepare my thesis on my own. Having completed my draft, I brought it to him. After he had read it, he gave me his opinion. "I liked it", he said, "with the exception of the first and the last chapters." I then revised these two chapters and made some changes in the middle, whereupon he told me to submit it formally, but to expand it a bit since the committee might consider it too short. When, after I had blown it up a bit, and it passed, he congratulated me and said, "Now you can cut it down and submit it for publication." When eventually published, my thesis occupied a total of twelve journal pages.

The custom at Penn was for the whole department to attend the thesis defense. In my case, the examiners included the Plato scholar Glenn R. Morrow, the medievalist Francis Clarke, the ethicist and historian of philosophy Elizabeth Flower, the aesthetics professor John Stokes Adams, and the logician R. M. Martin, aside from Goodman.

I do not remember much about the occasion, except that I was asked many questions by my examiners, and that Morrow, in particular, grilled me for quite a while.

Morrow, as chair of the department at the time, had called me in for a meeting before the defense to discuss my future plans, as was standard practice at Penn. He had been very cordial though austere and was, as I recall, somewhat unhappy at my expressing a positive interest in teaching. He responded by warning me about the difficulties of finding suitable teaching positions, and encouraging me to think of philosophy as a humanistic pursuit independent of employment, as exemplified by T. S. Eliot, the bank clerk, and Wallace Stevens, the insurance agent. He told me to look into job opportunities at the post office.

I surmise he must also have been skeptical of the nominalistic tendencies displayed in my thesis, for it was these tendencies that formed the focus of his hard questioning at my defense. I recall answering him, as well as the other examiners, rather bluntly and undiplomatically, and I left the examination room after the two hour ordeal, feeling I had been too sharp and curt in my replies, too lacking in due deference. After about half an hour, I was called back to be told I had passed and to be congratulated. Later, I was told by Betty Flower of an exchange between Morrow and Goodman during the interval before I was called back. Morrow had complained to Goodman that, in complete contrast to my own impression, I had been too diffident in my replies, to which Goodman had responded, "Don't worry. After a few months of teaching, he will become as pompous as any other academic."

Many years later, I told Goodman how much I admired his thesis advising–in particular, his having spoken to me about my thesis a total of three times during the whole process, thus expressing his confidence that I could do it all on my own. He refused my admiration, saying that he typically spent a lot more time with student theses but was gratified that, in my case, I had made so few demands on him. I am not at all sure that his self-estimate was accurate. In any case, his benign neglect of me during my thesis writing stage signified to me his trust that I could do the job. Such neglect provided, I suspect, greater support than the hand-holding that has, since that time, become so much more prevalent.

A final aspect of his support was the evident fact that he valued our opinions. One indication of this was his attitude toward our participation in discussions with visiting lecturers. After the visitor's presentation, there was a question and answer period during which Goodman himself took active part. He was often, apparently, disappointed that the assembled students had not sufficiently participated, and he continually urged us, before such sessions, to speak up. His attitude, I believe, was not that he just wanted more participation–fewer dead silences-during the discussion period, although that too must have played a part. My sense was that he considered our comments to be valuable; the way our minds bounced off a speaker's presentation providing worthwhile clues to the quality of the thesis presented.

Another indication of Goodman's valuing of our opinions was the fact that he invited our comments on his views. This is certainly common practice as a pedagogical device, where the teacher throws some thesis, whether he espouses it or not, into the classroom arena to invite students' responses and so sharpen their perceptions and argumentative skills. But Goodman invited our responses to some of his own serious views. One such occasion was his reading to our seminar his paper "Sense and Certainty", which was to be presented at a symposium with C. I. Lewis and Hans Reichenbach at a session of the American Philosophical Association in 1951.[8] It was clear to me this was no pedagogical exercise but a genuine effort to get our feedback to his arguments in advance of their formal presentation. But intended as such or not, the reading of his paper to our seminar had substantial pedagogical effect. It bolstered our own self-confidence not by superficial encouragement but by demonstrating that he believed we might have something valuable to say on his own work.

Goodman's support for his graduate students embarking on their careers took different forms in each case. For Chomsky, who was a member of our seminar, it took the form of recommending him to Harvard's Society of Fellows. In my own case, he recommended me for a Ford Fellowship during my last year of graduate study. He did this without telling me about it, so the first I heard was an official notice from the Dean. But, more importantly, Goodman's support took the critical form of getting me an interview with the Dean of the Harvard Graduate School of Education, Francis Keppel, an interview that proved a turning point in my life.

CHAPTER 4

FRANCIS KEPPEL AND THE HARVARD SCHOOL OF EDUCATION

Goodman had been in Cambridge and apparently went to see Keppel on his own to put my name forward for a new position as instructor in education. It was the summer of 1952, just after I had received my Ph.D. at Pennsylvania, and I had virtually no other prospects for an academic career. One evening in early July, I received a phone call at my apartment in New Jersey; it was Goodman, who told me to show up the day after next for an interview with Dean Keppel at the Harvard Graduate School of Education for an opening as instructor in education, with special reference to philosophy of education. I was flabbergasted, not having had any inkling of such an opening nor any hope of success at Harvard after the one live interview I had had early in the month, for a philosophy instructorship at Yale, an interview that had turned out an unmitigated disaster.

I had arrived in the morning at the office of the Yale department chairman, Charles Hendel, who briefed me about the position and ushered me into a nearby room where two members of the department were going to question me. They proceeded to grill me furiously about my dissertation for about half an hour when two other department members took their place and continued the grilling for the next half hour, to be replaced again by two others.

During a welcome interlude for lunch, I was able to have a relaxed and pleasant conversation with Carl G. Hempel, Goodman's old friend and at that time a member of the department. Right after lunch, however, I was again brought back to the interview room, where the intensive questioning resumed under the leadership of two new interviewers who were in turn replaced by two others after they had finished.

Of the series of questioners I recall only two, one being the brilliant logician John Myhill, the other the senior metaphysician, Paul Weiss—the former memorable to me because he left after a short bout of questioning, displeased, as I surmised, with the nominalistic line I had been taking, the latter because, after about only two minutes had passed, he declared, "This discussion is too technical for me", and left. The interview process continued in the same vein for the remainder of the afternoon, as I became more and more fatigued with each new set of questioners. The procedure reminded me of what I had read about Stalin's police interrogations in the Lubyanka prison, the object of which was to wear down the weary suspect by continued grilling at the hands of fresh teams of agents in unending supply.

No doubt this comparison was an overwrought product of my increasing fatigue and tension, which were finally eased at the end of the afternoon when the chairman arrived to take me back to his office. We chatted briefly about the position which, he told me, carried an annual pay of about $4,300 or $4,500, I can't remember which. When I asked him if there were any provision for low-cost faculty housing, he visibly bristled as he said no, and I realized I had overstepped some line. He then curtly offered to drive me to the train station. During the drive, nothing was said about when I might hear anything about the department's decision on the position. Finally, arriving at the terminal, I steeled myself to ask the question directly as to when I would be notified of the result of my interview, whereupon the chairman airily replied, "Oh, I thought you understood that we weren't interested in you."

After that disaster, the only opening I had vaguely heard of that summer was for an instructorship at Bucknell, which Goodman had vehemently vetoed when I had casually mentioned it to him. Given this background, my mixed feelings at receiving Goodman's telephone call are understandable.

My wife and I were both excited and apprehensive, but had no time to dwell on our mixed feelings since the time was so short. Arrangements had to be made for the trip to Cambridge and for a hotel room for the night following the interview. I booked a sleeper for the next evening, boarded the train and found it almost impossible to sleep, worried as I now had time to be, about the interview. I had had no training in education as a field of study, had no knowledge whatever about the Harvard School of Education, and had had exactly two previous contacts with a Dean during my entire undergraduate and graduate career, neither of them thrilling. Tossing and turning, I managed eventually to doze off for a few hours before the train pulled into South Station in Boston, whereupon I needed to dress and ready myself for the morning meeting with the Dean at his office in Lawrence Hall.

Lawrence Hall, which no longer exists, having burned down several years ago, was an old building, whose age showed on its face. It had much earlier housed the Lawrence Scientific School, where both C. S. Peirce and William James had had classes, and its basement hall, where I later gave lectures, had old wooden desks with names of many student generations boldly carved into the wood. I had little time to ponder the building, however, nervous as I was and as ready as I ever would be, to meet the Dean.

My preconception of what a Harvard Dean would be like imagined a solemn middle aged or elderly gentleman with a sober, unsmiling expression and a slightly forbidding air. Ushered into his office by Miss Fouhy, his secretary, I found myself instead face to face with a slender, relatively young man, who couldn't have been more than about ten or so years older than myself, and who welcomed me with a smile and a handshake. After we had introduced ourselves, his next words to me were, "Been to the can yet?" Completely baffled and sure I hadn't heard him right, I was too tongue-tied to respond, whereupon he put me at my ease and continued, "It's been a long trip from the Station, why don't we visit the men's room before our meeting?" So, with him leading the way, we trudged down the hall to the men's room, and continued to make small talk from adjoining urinals before washing our hands and returning to his office for the interview.

As I found out much later, this was vintage Keppel, puncturing all formality, undoing all preconceptions the visitor might have had, and disarming all preparations that might have been in the air. A paradigm case of this treatment, about which I heard many years afterward, was his reception of a well-known philosopher of education who was at that time a young education reporter for one of the news magazines. She had been assigned the job of interviewing the dynamic new Education dean of Harvard, good things about whom were already beginning to spread beyond the ivied walls.

Equipped with a serious professional attitude and a respectful mien, she had been ushered into Keppel's office, and invited to seat herself; she then opened her notebook, pencil poised, and prepared to record his replies to questions she had ready. But before she could start, Keppel exclaimed, "What interesting glasses you have on!" "What do you call them? Oh, harlequins?" (In fact, she had on blue tortoise-shell harlequin glasses.) When she stammered, "Yes, harlequins", Keppel added, "Do you mind if I try them on?" When she nodded her approval and wordlessly handed the glasses across Keppel's desk, he took them and put them on, calling to his secretary in the anteroom, "Miss Fouhy, would you please come in for a moment?" As soon as she stepped in, he turned toward her and asked, "How do I look?" After the somewhat bewildered laughs had subsided and the by-play was ended, the reporter retrieved her glasses, but by now all her solemn anticipation and preparatory attitudes had been shattered beyond repair. With all the formal stuffing knocked out of the interview framework, a new level of straight talk could proceed.

To return to my own interview session, when we reentered Keppel's office and seated ourselves across from each other at the long table that served as his desk, I awaited the first question he would be likely to ask me–something about my background or my graduate studies, or perhaps my fitness to teach education. Instead, he leaned back in his chair, put his feet up on the table, and proceeded to talk to me about the School of Education, when it had been founded, how it had fared, what problems it had encountered, what issues it had had to face in the recent past, where it was now going. He went on and on about the present state of the school, what his hopes were for its future, what its influence ought to be on the public education of children, and so forth. When, after about an hour, he had finished his comments without my having said a word, he asked me if I had any questions.

Hearing none, he continued, now telling me more about the job itself which would be in the area of philosophy, and would fit within his plan to enlarge the number of joint appointments, then quite small, and so to forge closer ties between the Graduate School of Education and the Graduate School of Arts and Sciences. In fact, support for this new appointment in philosophy had a counterpart in support for a new appointment in history, both underwritten by the Rockefeller Foundation, with the purpose of bringing new strength in the liberal arts disciplines to the field of education. Finally, I interposed a question: "Can you consider me a candidate for the job you describe, when in fact I have never had an Education course in my life?" "If you had had one," he replied, "We would not be able to hire you; the whole point of these two appointments is to bring scholars trained outside of education to the field of education so as to provide new perspectives on its work."

I protested that I would not, however, be able to lecture on education, having no substantive background in it, and that the only field I felt able to lecture on was philosophy. In lecturing on philosophy, did I think, he asked, that I might, on occasion be able to use educational matters as examples? This I told him I thought I could in fact do, whereupon he seemed satisfied.

He then asked whether I knew anyone in Cambridge. I told him I knew Noam Chomsky since we had both studied at Pennsylvania with Goodman, and Chomsky, nominated by Goodman for the Harvard Society of Fellows, had joined the Fellows a year or so earlier. "Why don't you go into the office next door, and give Chomsky a ring", he suggested, "ask him about Harvard and the School of Education, and discuss this position with him." He directed me to a vacant office, where I succeeded in calling my friend Noam.

I briefed him about the position in question and told him the Dean had suggested I ask him what he thought of Harvard. Knowing something of Chomsky's blunt and unconventional opinions, I had a vague idea that his reply would be quite different from what Keppel might have expected. As I thought, Chomsky had a number of unflattering comments to make about Harvard—roughly to the effect that it was a school populated by the rich and the powerful, full of self-important and pompous characters both propagating, and taken in by the Harvard myth. I said "thanks, and goodbye" to Chomsky, and returned to Keppel's office. "Well, what did your friend say?", he asked. I proceeded to tell him as accurately as I could, what Chomsky had said, not watering down any of his comments and waiting to see his reaction. Not turning a hair, Keppel said something like, "That's interesting", and coolly went on to conclude our interview. He told me he would be in touch soon, after the decision had been made about the job, and he said goodbye. It had been quite an experience for me. I knew I had been in the presence of an original—an extremely intelligent, extremely clever, and unusually insightful man. I had no idea whether I had made a good impression or not, nor any idea whether I would hear from him again.

I did hear from him, in a week or so, offering me the position and asking me to come to Cambridge again, to arrange details. When I arrived, Keppel was jovial, confirming my renewable appointment as instructor for the coming year, to begin in September, that is, in about six weeks. I had a hundred questions, but managed to ask only the first, that is, what did he want me to teach. In reply, he launched into another long oration, this time not about the history of the Graduate School, but about his conception of teaching within its purview, an oration which revealed his startlingly original ideas.

As a preliminary to his reply, he repeated what he had told me on my previous visit, that the School was dedicated to the improvement of free public education in America, and that he hoped I understood and accepted this direction as a background to my teaching. He also suggested, as a guideline, that I think of half my time as devoted to teaching and preparation for teaching, with the other half set aside for my free research and writing. He wanted me to understand that faculty members were all equal partners in the endeavor, that no distinction of rank was to interfere with our cordial relations, that we were to call each other by first names only, a practice he began then and there. Finally, as to what he wanted me to teach, he said he believed in the horse racing maxim, "We bet on the horse, not on the track." He wasn't going to tell me what to teach at all, but was going to leave the content of my teaching completely up to me, adding that he, as Dean, did not consider himself a scholar but as one whose job it was to facilitate the work of scholars.

With just a few weeks left before the beginning of the Fall semester, I felt a sense of panic looming at the prospect of making my first fateful teaching decisions. But I had not counted on Keppel's further surprises. He went on to say that I ought not start teaching at all in the coming term. I needed time to look around, get to talk to other members of the faculty, acquaint myself with the variety of courses and programs already in place and think about what I might like to teach in the later, Spring semester. I ought to begin, he suggested, by making appointments to talk individually with as many faculty colleagues as possible, and to discuss with them what they considered to be the role of philosophy at the Graduate School. Then I might be in a better position to reflect on the prospect of my teaching and to decide freely what I'd like to do.

This surprise was not merely *ad hoc*; it came out of a general conception which he had thought out and proceeded to explain to me as follows: The prevalent view, he said, was that the beginning instructor should have the most strenuous teaching load, and that, as he gained experience, his load should be progressively lightened until, with tenure and seniority, his load should be the lightest of all. This prevalent view, according to Keppel, was the exact opposite of the most desirable and rational arrangement. The beginning instructor, who still has to find himself, organize his ideas, plan his original research, and develop a personal style of teaching, needs the lightest load of all. As he continues to resolve these issues in later years, his load may be incrementally increased. The tenured senior professors ought, at their stage of teaching, to be presumed to have worked out all these problems already and should therefore have the heaviest load of all.

In line with this conception, he not only wanted me not to teach at all during my first semester, but proposed that I teach only one course during my second, a course of any type, of my own choosing. Moreover, since he recognized the special situation of those in the faculty of Education who had been trained primarily in the Arts and Science disciplines, and were accordingly identified intellectually with that faculty, he said he would try to arrange it so that my one beginning course would be cross-listed under both Faculties. If that were not immediately possible, he added, he would try to arrange it in subsequent semesters. In any case, my course would be open to students in Arts and Sciences as well as Education.

He warned me that the quality and preparation of students in Education had not been as good as those in Arts and Sciences, but that, however true that had been in the past, he was determined to change the situation. Therefore, since I was quite likely to have students of both sorts in my classes, I was to treat them exactly alike in the expectations I had of them, the standards they were to meet and the grades they were to be given. With equal treatment, the relevant levels of quality and preparation would be equalized, if not immediately, then over time. He wanted, above all, for there to be free student traffic in both directions—from Education to Arts and Sciences and vice versa.

Only equality of this sort would be capable of raising the academic level of Education from the lowly position it had occupied, to the level of Arts and Sciences. Ultimately, the raising of the prestige of Education would mean that it could compete more favorably than before with the Arts and Science disciplines for the ablest students in each undergraduate student cohort. Eventually, the teaching profession at large would also compete more favorably with other professions, such as medicine and law, for the best and brightest in each generation. With that, the meeting seemed to have reached a natural conclusion, but Keppel had one more bombshell to explode. Now that my appointment was confirmed, he had one last bit of counsel for me: he wanted me to think about what American education would look like in a decade or so and what I would like our School to be and do in the same period of time.

My head spinning, I left the meeting full of thoughts moving in several directions at once. I liked the fact that my charge was to learn about the School by conversations with the faculty and then to choose what to teach in an environment that bridged Arts and Sciences and Education. I also was thrilled that my duty was to uphold the standards of philosophical study and to do so without distinction between one group of students and another. But then apprehension set in at the thought of meeting all my new colleagues, most of them older than myself and senior to me in rank, and later on having to decide what to teach and how to teach it.

I needn't have been apprehensive about the meetings themselves for my new colleagues were all friendly and welcoming if not very helpful to me in deciding my teaching plan. Most had no clear idea what they expected philosophy to contribute, nor did they have any adequate conception of the subject. Some spoke vaguely about values, some invoked the name of Dewey, some indicated they identified the field with the history of educational practice, and so forth. I did meet many interesting and admirable scholars during the next semester, but the variety of opinions I encountered left me with the conviction I had best make up my own mind what I could properly teach within the scope of my new position and hope to learn more as I went along.

Eventually, I decided I would like to teach a seminar on ethics. My graduate concentration had been in philosophy of science and philosophy of language; my thesis had dealt with indirect discourse, a topic within the latter field. Although I had had a couple of courses in ethics and was familiar with the broad history of the area, I felt that I had not gained an understanding of the recent analytic approaches to the subject, and wanted to learn more. The best way to learn something is to try to teach it, I thought, so I gave Keppel my proposed seminar title, "Analysis of Ethical Discourse" and asked if it could be cross-listed in the Arts and Science department of philosophy. Although I'm sure Keppel did not have a real grasp of what I wanted to do, he was glad I had made a decision, and did in fact try to arrange for the course to be cross-listed, even though my position was not yet a joint appointment with the philosophy department as it in fact became many years later.

So I chose as my text W. Sellars' and J. Hospers' book, *Readings in Ethical Theory*[1] and decided that the students enrolled in my seminar and I would all learn ethics together. I planned to assign one of the papers in the book for each session, and to ask one of the students to report on the paper to start the discussion. I was nervous indeed as I entered the seminar room for the first time and even more so when I saw that the resident Harvard ethicist, Professor Henry Aiken, had decided to welcome me by visiting my first class.

I did in fact survive and managed to learn quite a bit that semester about analytic ethics. But what a star-studded group my seminar was, as I gradually kept finding out! It included, among others, Evelyn Masi whose *Journal of Philosophy* paper on C. I. Lewis I had criticized while I was still at Penn,[2] Ernest Gellner, visiting from the London School of Economics, a philosopher and anthropologist, Jonathan Cohen, visiting from Oxford, a philosopher of law and language, and Ronald Dworkin, later philosopher of jurisprudence–with occasional visits from Stanley Cavell, later Harvard philosopher of aesthetics, and Burton Dreben, later Harvard professor of logic. Had I known, to begin with, what a distinguished group this was, I would have been even more nervous than I in fact was.

It is interesting that this first class, sponsored by the Education School, had not enrolled one education student, all seminar members being philosophy scholars. I was somewhat worried that Keppel, who was paying my salary, would not take kindly to this arrangement. But I need not have worried. He wanted to draw the Education School and the School of Arts and Sciences closer together, and having me represent the former within the inner sanctum of the latter suited him fine. He was convinced that, over time, students from both schools would find their way to my classes, that philosophy itself would become a familiar Education subject and that Education students would find their way into classes taught by other members of the philosophy department. He hoped, as well, that Arts and Science students would begin to come to classes I would teach at the School of Education. Thus, he hoped to enrich the field of education by attracting serious philosophical attention to it and, at the same time, to improve the reputation with which the field was held by the general public.

What he could not have foreseen fully, although he may have had a hopeful glimmer of it, was the effect on me, trained as a philosopher, but teaching Education students, rather than primarily scholars planning to enter the ranks of philosophy instructors after earning their Ph.D.'s. Members of the latter group were already specialized in their backgrounds and career aspirations. They all had the usual, at least minimal, training in history of philosophy and the major branches of the subject, that is, metaphysics, epistemology, ethics and logic. Having spent years as students within the confines of graduate philosophy departments, they were now hoping to join such departments for the remainder of their working lives. Their intellectual worlds revolved mainly around the concerns of the profession, new professional research with which they were increasingly occupied as they were preparing their theses, luminaries of the profession determining the status of departments they hoped to join, positions that were open or likely to became vacant when they were ready to apply for them, the politics of the academic profession, especially as it related to philosophy, and so forth. These students were already socialized in the profession, they had so much in common that they could take a host of assumptions for granted in their mutual encounters, never having to explain themselves, at least on fundamentals. Such a student, entering a philosophy department as an instructor after graduation, would be again confined to the same professional arena, never having to talk about anything intellectually serious to anyone outside it for the rest of his career. To teach such students was to be an instrument of their socialization and to be at the same time, increasingly socialized in the same way oneself.

My teaching Education students as well as philosophy students put me in quite a different position. Education students came with a great variety of backgrounds and they were faced with a multiplicity of possible careers. Most had little or no acquaintance with the subject of philosophy, at least as understood in academic departments of the subject. Of those who were preparing to teach, some wanted to teach in elementary schools, some in secondary schools, some at universities. They were further divided by the subject matter of their intended teaching, whether arithmetic, social studies, or English, for example. Their projected researches and thesis studies often involved psychology or sociology, or pedagogical methods or history, and implied competence in experimental or statistical or questionnaire methods, or case study techniques. Moreover, all such students were involved in practical issues affecting the education system, as the large context within which their careers were to be situated.

As a result, these students shared very little in the way of a uniform background of education or aspiration and they did not offer a unified, let alone receptive audience for the traditional teaching of academic philosophy. Keppel had made it clear that I was free to teach this group whatever I wanted and however I decided to teach it. He was wise to delegate the problem to me since neither he nor anyone else had a firm tested solution for it.

I decided that I could not possibly teach materials drawn from this variegated Education world; I could only teach what I knew and that was philosophy. But since I expected my materials to have some resonance within my students' minds, I could not possibly teach in a pedantic or authoritarian way. My solution was, in effect, to teach what I knew, encouraging students to react, and listening carefully to how they were receiving whatever I had to say to them. My teaching was reflected back to me after making contact with their minds, and gave me the chance to understand how they saw it and what their operative presuppositions were. Over the course of time, my own teaching content was gradually and incrementally turned in the direction of such presuppositions, not either to confirm them or negate them but to employ them as means to better communication. In teaching a varied audience, my own conceptions about the power and resources of philosophy were considerably modified and the contact I made with students substantially and progressively increased.

One consequence of my being a philosopher of education at Harvard was the fact that the school catalogue, available to the public, identified me as such. Now, the public had conflicting ideas about philosophy. Some considered it an abstruse and impractical form of speculation of no earthly use. Others thought of it as the fount of all wisdom, with answers to all questions that might arise in human life. The first group did not concern me but the second made itself known by questions addressed to me about a variety of educational issues to which I certainly had no answer. I was invited to address PTA meetings about school organization, class size, child development and related topics.

One particular query drove me to seek an appointment with Keppel. The questioner was a mother who wanted my advice on the best methods of toilet training her child. Keppel was clear and firm. My job, he told me, was to teach philosophy, as I understood it, and not to concern myself with any other demands from the public. There are other people who can deal with matters of child development and school practice, he said, and I ought not to be diverted from the tasks I was trained to do in order to talk on such issues; I should turn down all invitations that might consume the valuable time dedicated to research and writing. Thus reassured, I was able to concentrate on my proper job.

I started my first proper lecture course in the School of Education dealing with topics in American philosophy, and in successive years taught epistemology, analysis of teaching, ethics and topics in history of philosophy, increasingly gratified by the positive responses of students who saw in the subject often more that I had anticipated, of relevance to their future work in Education. Several students, over the years, told me they found my Introduction to the Philosophy of Education course the most practical course they had ever taken, offering fundamental ways to think about knowledge, skill, moral judgment, teaching, and schooling. Coming to my classes without foreknowledge of the subject, they typically ended intrigued by it. I came to think of philosophy as analogous to chocolate; before tasting it, one had no prevision of its power, but having experienced it once, one often became addicted. My students, throughout my teaching career, were my natural allies. They, unlike most of my faculty colleagues in the School of Education, understood and properly appreciated the force of the subject I was hired to teach them.

If such appreciation was, indeed, a rarity in the School, it is even more of a rarity within the American population at large, where two typical responses to the subject prevail: "Philosophy? It's too deep for me!" is one response. "Philosophy? A totally useless subject!" is another. Both responses converge in the view that the abstract nature of the subject destroys its relevance to the real world, turning its devotees into absent-minded dreamers or fools. The picture is ancient, harking back to the Greek tale of Thales who, gazing at the stars, tumbled down a well.

This ancient picture no longer prevails in most other countries around the globe, where the subject is often integrated into the secondary school curriculum and thence filters into the public discourse. In the United States, however, philosophy is not an option in the curriculum of the secondary school nor is it typically a requirement even in college study plans. As a result, American academics and intellectuals have no exposure to the subject in their own educations and can hardly inject its concerns into public deliberations. Coupled with the practical and anti-theoretical temperament of American culture, this fact has created a vacuum filled by popular psychology and the social sciences whose pretensions to moral guidance and epistemological understanding are now generally taken for the real thing. Teaching philosophy is thus, at best, a matter of swimming against the mainstream, and it was therefore heartening to me to see enthusiastic responses to my teaching, in particular from Education students who were not typically preparing to teach the subject themselves.

My first lecture course of about 45 students, did, however, get me into hot water with the registrar's office at the School of Education. Having turned in my grade sheet for the course, I assumed my work was finished. About a week later, however, I received an irate phone call from Judson Shaplin, Associate Dean of the School. "What the hell do you think you're doing?", he demanded. Completely baffled, I asked "What are you talking about?", convinced there had been some serious but innocent misunderstanding. "I'm talking about your grade sheet!" he shouted. "What's wrong with it?", I asked. "You can't give over fifty percent of your students a grade of 'C',!" he shouted. "Oh yes, I can", I retorted, "that's what they deserved." "No, you can't," he thundered, and he clinched his case by saying he would refuse to register my grades until I composed a new grade sheet with different grades, which in the event I had to do.

It turned out there really was a misunderstanding behind this exchange. I had interpreted a "C" as an average grade, following my past experience in grading undergraduate students at Penn, where I had earlier served as a teaching fellow. The expectation for Harvard graduate students, I learned, was quite different. A "C" was below average, and had to be balanced by an "A" if it was to count as a grade credited toward a degree, while a "D" had to be balanced by two "A"s. In other words, a "B" was the lowest passing graduate grade, automatically credited toward a degree. Even a "B-" was considered somewhat shameful. My presumption that "B"s and "A"s were both honor grades, to be issued sparingly only upon evidence that a student's work was above average, was just mistaken.

Years later, the Education faculty issued qualitative guidelines for the assignment of grades, which went somewhat as follows: "B" represents the quality of work to be expected of a graduate student; "A" represents the quality of work that exceeds the normal expectation; "C" and "D" fall below such expectation—they are in effect not passing and cannot be credited without suitable balancing grades; "E" is a failing grade and cannot be counted at all toward a degree.

All this was before grade inflation really took hold and well before it began to recede somewhat in recent years. At its height, students felt they had failed if they received less than an "A", where grades were not altogether superseded by the "Pass-Fail" system, in which very few ever failed.

In 1972, with grade inflation rampant, I had taught two summer school courses at Harvard and left for a year's sabbatical at the Center for Advanced Study in the Behavioral Sciences in California, eager to get to work on two book manuscripts I was working on. The first mail I received on the morning of my arrival at the Center brought me a complaint from one of the summer school students who had taken both my courses. He had received an A in one of the courses and an A- in the other, and complained that he should have gotten an A in both. To bolster his claim, he enclosed the final examinations he had written for both courses, which I had graded. Having re-read his examinations, I wrote to him explaining that no final grades could be changed unless the instructor testifies that the grade was an error on his part. I further wrote to say that, having re-read his papers I had concluded that if there had been an error, it was in giving him an A in the one course, instead of an A-, so that if he insisted on my attempting to change his grades, I would recommend his getting an A- in both courses. I enclosed his examination papers and asked him to let me know if he wanted me to try for a change. Needless to say, I never heard from him again.

After my initial run-in with Shaplin, we enjoyed quite friendly relations until the end of his term at Harvard with Keppel's leaving in 1961. Shaplin had been, along with Robert Schaefer and Edward Kaelber, one of the excellent young administrators that Keppel had chosen as his close associates in running the Education School. With Keppel setting the tone, all succeeded in creating an atmosphere of collaboration, freedom, respect for research, and commitment to the study and improvement of education. This atmosphere was a product of the administrative personalities at the School, no doubt, but also a result of their constant effort to promote opportunities for informal interaction among faculty and students.

One such opportunity was provided by the daily tea instituted by Keppel. Every weekday at 3:00 p.m., a tea was provided in one of the School's rooms, to which all were invited. Students and faculty members would drop in any time their schedules permitted, and these occasions provided a pleasant interval that helped to break down formal barriers and raise morale. Occasionally, too, faculty retreats were arranged where faculty members would convene in a locale away from Harvard to discuss School problems and plans for the future.

An occasion of this sort provided an amusing opportunity for me to get to know Shaplin better. A faculty retreat was arranged for two or three days at an inn in Chatham, on Cape Cod. This, as I recall, took a bit of doing, since some of the inns our administrators originally inquired into objected to the likely presence of Jews in our party. Eventually, however, a suitable place for our retreat was found. Most of the participants arrived at the inn on the eve of the first day of meetings. We had an opportunity to settle into our rooms, meet for dinner and engage in conversation in preparation for our meetings the next day. When we awoke early next morning, ready to work, we got unexpected news. Kaelber, who had been leery of the weather, had walked down to the Coast Guard Station early and been warned of a serious hurricane bearing down on New England. The best advice we could get was that, although it was still several hours away, it would be a good idea to leave as soon as possible.

We all scrambled to pack up, check out, and get into our cars right away. Without having held even a single meeting, we started our drives back to Cambridge. Shaplin came in my car for the drive back, and we raced the hurricane for the entire trip, discussing the School and the weather all the while. Arriving in Cambridge, we stopped at a store to buy a heater and food warmer that Shaplin needed, before I dropped him off at his house. I returned home just minutes before the hurricane struck, a hurricane that cemented my friendship with Shaplin.

In the philosophy department, as distinct from the Education School, I taught some of the standard courses to a more homogeneous philosophy audience, e.g., philosophy of science, philosophy of language, pragmatism, but as the years progressed, these classes also had Education students, while my Education School courses began to have some philosophy students as well. In this respect, Keppel's idea became more and more operational; the mixture of philosophy and Education audiences in my classes continued to grow, and I learned correspondingly how, without watering down the material, to communicate its content to a general audience.

During my first year on the job, I spent a good deal of time, following Keppel's plan, talking to various Education colleagues, as I've already mentioned. Keppel, however, had still another ace up his sleeve in his effort to socialize me and several other new faculty members into the sphere of Education. He told us that he had decided to reform the School of Education catalogue, which seemed to him too scattered and altogether lacking any evident principles of organization. Indeed, it listed seventeen different fields of study without exhibiting any rationale for the conglomeration. In order to get the reform under way, he appointed a committee, chaired by a senior faculty member, the statistician Phillip Rulon, who had served as Acting Dean before Keppel's own appointment. As members of this committee, he named myself and a handful of other newcomers to the School.

Rulon was a distinguished and creative statistician who had pioneered multiple variation analysis. He had trained a number of graduate students who went on to notable careers in the field of statistics. But, unlike his Arts and Science friend and colleague, the statistician Frederick Mosteller, he found it very difficult to produce a book. Keppel had been using all sorts of incentives and devices to stimulate Rulon to write, but to no avail. He was an enormously influential teacher and graduate advisor, and produced important research papers but the comprehensive book eluded him.

A physically impressive man, he was tall and well-built, with a completely shaved head, typically wearing tall boots, and normally arriving at his office on a motorcycle. It was rumored that, having often consulted on statistical matters to the U.S. Air Force, he refused to take a money payment for his services, requesting instead a half-hour's use of an Air Force jet which he piloted solo. On his motorcycle one morning, stopped at a red light, he was hit by a car, which gave him only a scrape but damaged his cycle slightly. Thereafter, he was heard to remark, with a mixture of surprise and indignation, "You know, those motorcycles aren't safe."

At a special dinner for Rulon, shortly before his retirement, he received various laudatory comments from his students and colleagues, including Mosteller. In response, he related a story that seemed to epitomize his statistical credo and his abhorrence of certainty: One night, he and Mosteller had been talking about their researches, and drinking, into the early hours when the topic turned to death and the conflicting criteria that might conceivably be taken to certify a person's having died. They discussed the question until the wee hours in its epistemic ramifications until, not being able to reach a conclusion, they decided to call the morgue. Their call was passed along from one official to another until they were in phone contact with the attendant who actually received and stored the corpse. "How do you know when someone is dead?", they asked. To which the attendant replied, "When a body is delivered to me with a tag on it that says he's dead, he's dead."

Under Rulon's efficient chairmanship, our committee met two or three mornings a week for about a semester, each session devoted to hearing testimony from a faculty member working in one or another of the programs enumerated in the catalogue. Fortified by strong coffee brewed in Rulon's favorite Chemex, we heard such testimony and discussed it with the faculty witness, and amongst ourselves, while Rulon conducted the questioning and took copious notes, using his skill at shorthand, of which he was inordinately proud. These sessions were all quite serious, but some had their funny moments. One outstanding example was the day Robert Ulich was the witness. Ulich was the senior resident philosopher and historian of education, who had been invited by President Conant to Harvard from Germany and who had managed to get to Harvard just before the Second World War. A classically educated scholar steeped in European civilization, he taught, aside from philosophy of education, also comparative education, history of universities and, on occasion, religious education. Ulich and Rulon were none too fond of each other, at opposite ends of the academic and cultural spectrum as they were. Here was Ulich, the gentle classical humanist, being grilled about the content of his classes by Rulon, the gruff American statistician. "What do you teach in your basic course?" came the first question to Ulich. "I teach the Renaissance", was the reply. "The Renaissance, eh?", said Rulon, and duly recorded the reply in his shorthand.

Over the course of the semester, our committee learned everything about the internal content of the School's teaching, direct from the mouths of those responsible for carrying it forward. We were properly introduced to the intellectual workings of the School in the company of one another and so were formed into a knowledgeable young cadre who had learned a common language for discussing the operations and tasks that the School had set for itself. And, as it happened, Keppel's idea had an extrinsic result as well. Rulon's committee came up, at the end, with a plan to overhaul the School catalogue, eliminating the various separate rubrics with which we had started, and concluding with a neat tripartite division into Humanities, Social Sciences and Clinical Practice, which gave an immediate integrated picture of what the School was about.

During my first year, while I was still meeting various colleagues to share ideas about the School, I would occasionally meet Keppel in the corridor in front of his office. He would always greet me warmly and invite me into his office. I felt diffident about taking his time, but he always had time for his faculty, especially the new members. The table in his office, which served as his desk, was always piled high with papers at one end. As in my first interview with him, he leaned back in his chair and swung his feet up on the table—a man who apparently had all the leisure in the world. But now, instead of talking to me about the School, he would ask me what I was doing. He wanted to hear about papers I was writing, books I was planning, conferences I would be attending, seminars I was teaching and so forth. This was a most unusual dean, who knew more than most what was going on in the minds and aspirations of his faculty.

On occasion, he would ask my thoughts about what the School should be doing about this or that matter and we would discuss the issue. But I pretty soon noticed a peculiar habit of his. He would think of some new program or initiative he considered worth trying and, telling me about it, would preface his account by saying, "As you suggested ...", attributing to me the credit which belonged to himself for coming up with the idea. The first couple of times, I was too taken aback and too rattled to do anything about it. But the next time, I called him on it: "Frank", I said, "that was your idea, not mine. A good idea but the credit for it is yours." He was more cautious about using his ploy with me thereafter. When, in 1961, he was appointed U.S. Commissioner of Education and had to resign his Deanship, I was truly sorry to see him go.

One story, about Keppel's first day on the job in Washington, filtered back to us in Cambridge. Possibly apocryphal but surely in character, this story was related to me by the mathematician, Edwin Moise, and was meant to illustrate Keppel's shrewdness. Presumably, as he sat down at his new desk for the first time, he was presented by some bureaucrat with a large stack of official letters and documents and requested to sign them as they were urgently needed. Keppel's response went something like this: "Oh, the pace here in Washington is so much faster than what I've been used to in Cambridge. I couldn't possibly sign anything so soon." And, after the bureaucrat had left the room, Keppel tossed the whole stack into the wastebasket.

His career as Commissioner had been going well until he ordered Chicago to desegregate its schools, in accordance with President Johnson's policy. When Chicago refused after repeated urgings, Keppel ordered federal funds to be withheld from that city, whereupon Mayor Daley phoned Johnson in a fury and demanded Keppel's ouster. A most honorable exit, followed by Keppel's appointment to head General Learning, a publishing firm, whose job it was to produce educational materials. The company, however, was not a success and was sold about ten years later.

In the late seventies, Keppel returned to the School of Education faculty as a lecturer, and I had the great pleasure of resuming our association and serving with him on some committees of the School.

Before he left, Keppel had put me forward for promotion to a tenure professorship and I was duly promoted, just before his departure in 1961, to the rank of Professor of Education. On my copy of a group picture of the senior faculty taken at that time, he wrote, "Your appointment gives me a special pride and happy memory."

His successor as Dean, Theodore Sizer, promoted me to the position of Victor S. Thomas Professor of Education and Philosophy, in which position I regularly taught from 1964 to 1992, both in the School of Education and in Arts and Sciences. Sizer also appointed me Chairman of a committee to report on the graduate study of education at Harvard and other research universities. Our report, *The Graduate Study of Education*, published in 1966,[3] developed several new initiatives and was well received until the general downturn with the Vietnam War, the trouble in the cities and the resultant decrease of funding for education. Educated as a philosopher, I now found myself regarded locally as somewhat of an advisor on educational policy.

CHAPTER 5

SOME EDUCATION COLLEAGUES

During my first few years at the School of Education, I made the acquaintance of many unusual personalities, of whom Rulon was one. I want to sketch some of these in what follows. Robert Ulich I have already briefly mentioned. He was the senior philosopher in whose area I was hired to teach. I have already described him as a classical humanist who had come to Harvard from Germany at President Conant's invitation. His work was historical and comparative in spirit and he was dedicated to promoting a vision of democratic education that was more classical in outline than was Dewey's, although he was not by any means a staunch opponent of Dewey's. His background and mine were quite different. My philosophical training had been primarily analytical and centered on logic and epistemology rather than history, and my sympathies were with Dewey's pragmatism. He was pleased, when I was hired, to learn that I had had a strong Jewish education, having studied at the Jewish Theological Seminary, for it gave me entry into a classical religious tradition. His own tradition was Christian but he had an extremely broad conception of religion, being drawn, among other things, to Asian thought, and developing a course in the history of religious education, which he taught for some years at the Harvard Divinity School.

He had been described by some as an existentialist but did not much like to be labeled a member of a particular school of thought. He felt close to Jaspers, but had little use for Heidegger because of his Nazi affiliations and shameful treatment of his Jewish teacher, the philosopher Husserl. Ulich had had one student, Frederic Lilge, eminent scholar of comparative education, whose doctoral thesis, written under Ulich's supervision, was entitled *The Abuse of Learning*,[1] and recounted the unconscionable capitulation of German academics to Nazism when Hitler took over the universities.

A kind man, Ulich received me graciously when I first met him; he had been abroad during my appointment and had returned soon thereafter. He invited me and the historian Bernard Bailyn, who had been appointed at the same time as I under the same Rockefeller grant, to offer a joint seminar with him on the topic of leadership, an enterprise which encouraged us to try to communicate across the borders separating our three different backgrounds.

During that initial year, also, Ulich formed a small discussion group which met weekly at his house on Shepard Street, and which included Bailyn, myself and a few others. Keppel may have been behind the idea, I surmise, thinking it would be a fine way to get us together in a collegial spirit. In any case, the discussion group was charming. We met in the afternoon in Ulich's living room and were served green tea in delicate cups while we discussed whatever topic had been chosen for the session. Ulich's small dog would join us and listen to our discussion or wander about and return as the spirit moved him. As I recall, I was not very deferential in my contributions to these discussions, but stated my views or reactions to others' comments rather bluntly, since that was the way I had been trained. On occasion, I could sense from his expression that Ulich was a bit concerned by my manner or my views, but he never in any way departed from his usual gracious demeanor. Occasionally the conversation would drift to European political developments and it was quite clear he was seriously worried about the situation in Germany as well as neighboring countries. All in all, a pleasant series of occasions, presided over by an admirable man. When Ulich retired, years later, he was given a small office in the School of Education and devoted himself to his writing. He wrote poetry and planned to write a novel set during the time of the Reformation. He and I would often stroll around the Radcliffe yard during the early afternoon and discuss the events of the day. I was much troubled about the state of the university during the difficult sixties, but I remember his reassuring me. "You know, Israel", he'd say, "the university has survived worse times during its history, and it will survive this time too". Eventually, he returned to Europe, saying, "America is no place to grow old in".

One of the first senior faculty members at the School of Education whom I met was the anthropologist John Whiting, who directed The Laboratory of Human Development at the School. He was a promoter of cross-cultural studies of child rearing, and received me cordially as one involved with philosophy of science, which he deemed an ally of his investigations. Indeed, he was preparing a manual for field studies of other cultures, which he asked me to look over and give him some reactions. A genial man, he and his wife, the equally genial anthropologist Beatrice Whiting, made an excellent team at the School, bringing the perspectives of other cultures to bear on American educational ideas and practice. Much of the educational training of our students had struck me as too local and foreshortened, and the correctives of anthropology and history were partners in broadening its scope. Both served also to stimulate philosophical reflection on education by casting doubt on the natural presumptions that what we do in our own locality and in our own period is the only and the best way to do things.

An interesting project that brought this lesson home in a striking way was a file of cross-cultural practices compiled from about one hundred and fifty cultures, for which empirical data had been collected by the fifties. In those pre-computer days, the information was sorted on index cards and located in one of the offices of the anthropology department, as I recall. The information was cross-referenced, so that the variation among the hundred and fifty cultures on numerous dimensions could be easily located. I remember looking up "breakfast" to find the hundred and fifty ways people ate their first meal of the day. Needless to say, this experiment was an eye-opener and an immediate antidote to any lingering ethnocentrism I had harbored. Of course, Whiting's main idea was not simply to display cultural variation but to propound general hypotheses about human conduct that would have to run the gauntlet of empirical variation in order to stand up. The found variations constituted, in effect, a library of counterexamples to any theoretical idea about human practices that an investigator might think promising.

One of the things I found most intriguing about Whiting was his speculative verve, which made his conversations sparkle with unusual new suggestions of explanations for social phenomena of one or another sort. This characteristic showed itself in ordinary conversations, in academic meetings, and in the social evenings occasionally hosted by John and Beatrice at their home. But I discovered after a while that it was the key to his great success as a graduate advisor responsible for helping to guide students through the thesis stage of their studies.

The laboratory which he directed was housed in a small wooden Cambridge house with a wonderful old kitchen, the centerpiece of which was an old-fashioned stove. I had heard rumors of what went on in that kitchen. Whiting required all of his doctoral students to have lunch together daily in that room, the lunch being a hot meal prepared by one of the students themselves. Next to the table at which they ate stood a blackboard. The one rule for the lunch was that shop had to be talked. So along with the communal meal, which symbolized and celebrated the common effort at understanding human life, went the intense exploratory discussions, by all present, proposing their suggestions, innovations, arguments, criticisms, hypotheses, etc. out of which emerged a wide variety of research ideas leading to doctoral dissertations. Over the whole lunch and discussion session, John Whiting presided, participating fully in all the conversations, occasionally punctuated by someone's drawing or jotting something on the blackboard.

Having heard about these "kitchen seminars", I asked Whiting if I could visit one. He told me it was a strict rule that no visitors were allowed. I imagined that the reason was the need to keep the group concentrated on their continuing discussions after they had developed their own little culture. This, I thought, was certainly an excellent reason but I still had the intense desire to see at least one session in action. So, on every occasion when Whiting and I met, I raised the question again, making it clear that I would remain absolutely quiet and would only need to visit just the once. Eventually, he acceded to my request and I was able to verify for myself that what I had heard about the sessions beforehand was correct. The atmosphere was warm and friendly, but serious. The discussion got off to a good start before the group had completely finished their tasty lunch. Whiting did not steer the discussion but pitched into it as an equal, and the ideas of the group flowed freely.

After the session was over, I had occasion to ask Whiting how he had fixed on that format. Some of the background, as I had already surmised, was bread and butter for anthropologists: the need to create an intimate face-to-face culture with its continuing activities governed by common values and rules. But what I had not anticipated was Whiting's personal affinity for the discussion style.

He told me that he had all his life learned less from writing than from talk. Throughout his own advanced studies, he said, he had almost never taken any notes, relying on his memory for what was said to convey the points being made. Similarly, his own ideas developed first in conversation and discussion before eventually being put down on paper. The particular format of his kitchen seminars was thus a common ground where his personal style and his professional predilections came happily together. This confluence of style and professional work is too rare in teaching but a circumstance supremely desirable. It is of interest to note that, after Whiting left the directorship of his laboratory, no successor was able to recreate the special ambiance that he had fostered. Later directors had to find other ways to carry on the work.

Whiting had a special influence on the design of Larsen Hall, built in 1965, several years after my visit to his seminar. He proposed that each floor of the building be devoted to a given discipline or area of study, and that each floor have a kitchen. This kitchen he designated the "watering hole" for denizens of the area, who would be drawn to it as a cultural hub. Whiting's anthropological habits of mind also had an influence on doctoral examinations both in the Arts and Science department of anthropology and in the Education School. He had decided that the practices surrounding final doctoral examinations were too dry and formal. They lacked ritual to mark the significance of the occasion. Whiting thus inaugurated festive rites to accompany the candidate across the doctoral divide. Whiting's students progressed from meal to meal: from communal eating at the kitchen seminars, to eating at the anthropology watering hole between classes and seminar, to the ritual feast as a newly minted Doctor.

Finally, I need to mention I. A. Richards, the eminent literary critic and inventor, with Ogden, of Basic English as well as co-author, also with Ogden, of *The Meaning of Meaning*.[2] I had, of course, known of Richards before I came to Harvard and had also read some of his work with appreciation. What I had not known was that, coming from England as a Harvard University Professor, he could have had his choice as to which graduate school to take as his home base. Despite his fame and authority as a literary luminary, he chose the Graduate School of Education in this capacity, because of his overriding interest in world education.

I was overjoyed to find him a colleague because of my own interest in philosophy of language, and had imagined that, with him and with John Carroll, another senior Education colleague who had authored *The Study of Language*,[3] I might find a common intellectual bond. As it happened, the three of us did in fact interact to a degree but did not form a close alliance. Carroll and I co-taught an Education course on research in education which centered more on logic and methodology than on the special province of language. Richards came to my philosophy seminar, at my invitation, to engage in discussion of a well-known paper of his on the topic of emotive meaning. But although we recognized a mutual interest in language, his work engaged a more practical level than mine, which was more theoretical in its scope.

Despite his far-flung reputation and indeed celebrity status in academic and literary circles, he was exceedingly decent and low-key in his dealings with students and colleagues and he treated me with great friendliness. I remember with pleasure the evening my wife and I spent at his home, a special occasion to which he and Mrs. Richards had also invited T. North Whitehead. When I was introduced to him as a new philosophy instructor, he replied, "My father was a philosopher", referring, as I soon realized, to the celebrated Alfred North Whitehead, author, with Bertrand Russell, of the *Principia Mathematica*,[4] which had inaugurated a new era in twentieth century philosophy by its systematic development of modern symbolic logic.

Alfred North Whitehead had joined the Harvard faculty of philosophy in 1927 and was thereafter a well-known local figure, who gained a worldwide influence not only through his path-breaking work in mathematics but also through his systematic process philosophy, his books on the history of science and his essays on various aspects of education. I digress here in order to recount my later contact with his legacy, a contact which illustrates my initial theme of the amnesia of universities. Some eighteen or twenty years after my encounter with T. North Whitehead at the Richardses, I chaired a faculty committee at the School of Education which produced the report earlier mentioned, entitled *The Graduate Study of Education*. One recommendation of this report was that the School establish a post-doctoral program for senior educators to be called "The Whitehead Fellowship Program". The faculty voted this program into being but shortly before it was begun, I received a request from some office in the central administration to answer the question, "Who was Whitehead, and what did he do?" I was asked, further, to write a brief justifying naming a Harvard program after him. In my naiveté, I had supposed the answers to these requests were obvious, especially in an institution where Whitehead had served as a distinguished University Professor from the twenties to the end of his life. But I had not counted on university amnesia.

To return to Richards, I had mentioned that he had strong practical interests in education. These he cultivated in an initiative of his design, Language Research Inc., in which, along with a collaborator, Miss Christine Gibson, he produced a series of language teaching books for French, Spanish, German, Italian, Hebrew and Russian, using pictures as the medium of education. These books, using stick figures to represent various objects and situations, were supposed to facilitate the student's grasp of the meanings of words by pointing them toward the referents of the relevant stick figures.

Richards had an overriding faith in the growth of potentially harmonious relations among peoples by the mutual learning of their different languages. The idea of Basic English had itself been powered by this faith, Basic, as it came to be called, providing a nuclear English vocabulary suitable for defining or paraphrasing the full range of English locutions and so providing a much easier entry into the corpus of English works than the usual route of tackling the lexicon whole. Of course, this use of Basic had to be supplemented by the work of providing paraphrases of complex writings into its cramped vocabulary. And this too Richards attempted, producing *The Republic of Plato: A New Version Founded on Basic English (1942)*.[5]

That his practical educational imagination was always active may be illustrated by two other ideas that he came up with. Both were rooted in his awareness that small, poor, or isolated communities could not count on there being full scale schools available to their children. The first idea was directed to countries in Africa, where isolated villages lacked schools or adequate faculties. This idea was to use skywriting as an educational tool. A plane would appear overhead at a given time for a given village or group of adjacent villages, and would write the letter "A" with white smoke against the blue sky. At the return of the plane to the same location the next day it would write "B" and so forth. Meantime, it would be doing comparable writing for distant villages. An ingenious and romantic vision; I do not believe it was ever carried into effect.

The second idea was indeed realized. The problem to be solved here was the scarcity of trained teachers for foreign languages and for other subjects as well. Richards' plan was to design a special school bus with enough desks and headphones to accommodate a classful of children, then to provide sequential series of tapes in selected subjects, one tape per lesson. On an appointed day and hour, the bus would arrive at a given school and park on its grounds. The relevant class would be ushered into the bus, and each child would don a headphone. The driver, who had no need to teach or to understand the lessons in question, would then activate the proper tape for the day, so that the class would be hearing it simultaneously, but individually, and be able to do whatever exercises were called for. The bus, built as Richards had imagined it, was a familiar local sight for a while, parked at the Education School's lot at the start of the day, before beginning its circuit riding in Arlington in the interests of education. It is of a piece with these ideas that Richards was fascinated by, and heavily involved in the language teaching problems in foreign countries worldwide, particularly in India, with its plethora of languages, where English served as common denominator, and China with its variant dialects, non-alphabetic orthography and tonal system.

Such practical interests, intense as they were, by no means monopolized Richards' mind. Aside from his work on the Republic, he also, years after I first met him, wrote *Speculative Instruments*,[6] *The Screens and Other Poems*,[7] *Tomorrow Morning, Faustus! An Infernal Comedy*,[8] and other works. To speak of Richards as in love with words is an understatement. On at least one occasion, invited to give a formal lecture, he strode to the podium, took a slim volume from his pocket and announced that he would rather read poetry than lecture, proceeding to read from the volume, to the delight of the audience. One occasion I recall with particular pleasure is a lecture of Richards at the School of Education, entitled "How does a poem know when it's finished"? He projected a poem of his in manuscript upon a large screen, with numerous changes marked throughout, and discussed with the audience possible rationales for the changes. The title question was the culmination. Why not, after all, go on making changes without a stop? An intriguing question!

It is pertinent to note how well Richards exemplified the role of the university scholar so strikingly set forth in Alfred North Whitehead's celebrated essay, "The

Rhythm of Education".[9] Having been captured by the romance of particular subjects at the initial elementary encounter, and entered into their detail and fine structure in the secondary phase, the student enters the final, or university phase, the one in which general relations are the focus, in which ideas are thrown into new combinations and produce new patterns of thought and life. Richards' interests in the literary and the technological, the speculative and the practical, the philosophical and the educational, the local as well as worldwide human situations, patently express the university ideal.

CHAPTER 6

SOME LONDON PHILOSOPHERS

For the academic year 1958–59, I was awarded a Guggenheim Fellowship which I decided to spend in London. After receiving notification of my award, I got cold feet and planned to ask the Guggenheim Foundation if I could change my proposal so as to allow me to spend my fellowship time in the United States. Having told my friend, Maurice Mandelbaum, a philosopher at that time serving on one of the main Guggenheim committees, of my plan, I awaited the result of his consultation with his colleagues. Briefly the response to my idea was, flatly, no. I could go to England as originally proposed and keep the fellowship, or stay in the U.S. and forfeit it. To this day I am totally grateful to Mandelbaum and his colleagues for their wisdom in rejecting my idea.

Thus, my wife and I and our two small children, ages 7 and 4, arrived in London in the Fall of 1958 and stayed until the Spring of 1959, when we left for several weeks of traveling on the continent before returning home. Before coming to London, I had written to Ernest Gellner, of the London School of Economics, of my prospective stay. He had been a member of my first seminar at Harvard and he now responded warmly to the news of my coming visit, and invited me to join the common room of L.S.E. I was very glad to learn also that the reading room of the L.S.E. library allowed smoking, since I was at that time a regular pipe smoker.

This fact, as it happened, gave me a new occasion to realize the peculiar nature of university communities. As I was working in the reading room one day and ran out of matches, I asked a nearby reader if he could let me have a match. By our accents, we recognized that we were both American. Introducing ourselves, we found that he, Louis Hartz, and I were both faculty members at Harvard and, further, that his office at Littauer and mine at Lawrence Hall were a stone's throw from one another. To top it all, his main work was in political theory and he had a broad general interest in philosophy. We had never met at Harvard but the need for a match brought us together in London, where we enjoyed getting together on several occasions during our time there.

In London, I divided my work time between L.S.E. and the Reading Room of the British Museum, where Karl Marx had written *Das Kapital*. My first letter to Dean Keppel after I was settled in told him that I was writing from this reading room, possibly seated in the same chair where sat the bottom that shook the world. At any rate, in this room I devoted myself to writing my book *The Language of Education*[1] while at L.S.E. I began working on my *Beyond the Letter*, not published until 1979,[2] and continued working on my *Anatomy of Inquiry*[3] and some other problems in philosophy of science that engaged me, especially after my contacts with the Popper group.

Karl Popper was the senior professor in philosophy of science at that time at L.S.E. and among his disciples were John Watkins, Joseph Agassi, J. O. Wisdom, William W. Bartley III (who had been an undergraduate student of mine at Harvard) and several others. Imre Lakatos and Paul Feyerabend had not yet come at that time, but became well known members of the Popper group later on. L.S.E. was a very lively place, and had several other important scholars, such as Morris Ginsberg and Alasdair McIntyre; Hartz also was active there. Students from various parts of the Commonwealth came to L.S.E. to study and it exuded a cosmopolitan atmosphere in which diverse philosophical views and political doctrines were in constant debate.

Gellner was kind enough to invite me, soon after my arrival, to a lunch at the common room where he introduced me to Watkins and to Agassi. We had a pleasant time together and Watkins, a friendly and charming person, invited me to attend one of the renowned Popper seminars at a forthcoming session. After the others had left, Agassi and I remained, having gotten into a debate on induction, on which I had recently published a paper he disapproved of. Agassi was a bright and an unusually aggressive debater and the result was that our extended argument lasted for several hours, well into the afternoon.

As I recall, Agassi was incredulous that my article had favorable things to say about Goodman's new riddle of induction and that in our discussion I had also revealed my interest in Hempel's paradoxes of comfirmation. Following Popper, Agassi had no use for induction at all, and had a low opinion, verging on an ill-disguised contempt, for Hempel's work. I found out only later how deep these attitudes were in the Popper seminar. On this occasion, we argued to a draw when we finally agreed to call a halt. As an amusing finale to Agassi's remarks, he told me he had recently met and had an extended argument with Chomsky who, he complained indignantly, was an unusually aggressive debater. Apropos of what, I don't know, unless perhaps as an oblique defense of his own abrasive style, Agassi told me the following story: An itinerant Jewish preacher in Eastern Europe arrived in a small village and was welcomed by one of the residents of the village. The preacher asked where the leader of the village was. The resident agreed to take him to see the leader. Having been introduced to him, the preacher asked, "Are you indeed the leader of this village?" Receiving an affirmative reply, the preacher raised his fist and gave the leader a resounding blow to the face. To the bewilderment of the leader and all the onlookers, the preacher explained, "That was just to show you I am no flatterer."

I was very pleased indeed to have been invited to attend one of Popper's seminars. I knew very little about Popper as a person, except for occasional rumors which emphasized his irascibility and his conservatism. Before going to London, I had made a point of reading Popper's celebrated volume, *The Open Society and its Enemies*,[4] as well as his smaller *The Poverty of Historicism*,[5] and became a genuine admirer of his thought. I was impressed by his clear and blunt style, the breadth of his learning, and his fervently democratic allegiances.

Before going to his seminar, I had had the further opportunity to attend his Presidential address at the Aristotelian Society in the Fall of that year. His address was entitled "Back to the Pre-Socratics"[6] and I still consider it a brilliantly provocative re-reading of the origins of scientific thinking. The discussion following his address was lively and Popper gave direct answers to his questioners, occasionally, however, telling them to think further about the issue after which, he said, they would realize his views were correct. This sort of reply struck me as patronizing and dogmatic and, as a matter of fact, later on during the discussion period I heard a noisy shuffling of chairs at the back of the room, and a whole row of listeners arose and left, making no effort to do so quietly. When the session was over, I asked a philosopher seated near me what the commotion was. He explained that the group that left was comprised of Professor A. J. Ayer and a number of his students, the likely target of whose protest was not so much Popper himself as the questioners who, in the protesters' opinion, had pulled their punches and failed to respond vigorously to his dogmatic replies, showing him undue deference. This public protest struck me at the time as marking a real difference from American practice where such action would be unimaginable at a formal philosophical meeting.

In any case, this episode gave me a little pause before my first visit to the Popper seminar, but not much. I still came with a high degree of admiration for Popper and looked forward to seeing the well-known seminar in action, resolved to be an observer merely and not to say anything. I came in early and seated myself in the second row, which like the first, encircled the oblong wooden table. About fifteen or twenty persons occupied the seats, with Popper and Watkins at the head. I had no advance idea of the format but quickly realized that Watkins was to present a paper. What I had no prior inkling of was what presenting a paper meant in this seminar.

Watkins, the heir apparent to Popper and, in fact, later his successor in Popper's chair at the L.S.E., began by reading the first sentence or two of his draft, whereupon Popper stopped him and proceeded to criticize what he had just read. When he had finished reformulating Watkins' sentences to his own satisfaction, he motioned to him to continue. Watkins proceeded but he had not gone very far before being once more silenced by Popper, who again launched a number of criticisms of Watkins' preceding words. When he had done, Watkins gamely went on, again to be bidden to wait, while Popper once again proceeded to reformulate what we had just heard. I do not now recall the content of Watkins' paper nor the gist of Popper's remarks. I can, however, record my memory that these latter remarks were quite uneven in their scope and quality, some being relevant to the main thesis being offered, others being subsidiary to the central theme, still others being in effect repunctuations of Watkins' sentences. The speaker had had no chance to expound his basic thesis or to develop it in a coherent fashion, his presentation being chopped up as it emerged, reaching the audience in bits and pieces the full bearing of which could hardly be divined by this process.

Occasionally, there was room for some general discussion pertinent to something that had been said but the spirit of the discussion was disturbingly partisan. In particular, when the name of Carnap happened to come up, the whole group booed or hissed, and when anyone criticized Carnap, they cheered. Carnap's efforts at system construction were occasionally referred to, eliciting general booing, and when someone loudly declared that Carnap's constructions all collapsed, loud cheers ensued. To give Popper his due, he intervened to restrain one such demonstration, but without complete success.

About a half hour into the session, despite my resolution to remain silent, I asked a question of information, to which Popper replied but, despite my having been introduced by name at the beginning of the seminar, he addressed me now as "Mr. Shimony". Had I been more alert at the time, I would have recognized that slip as a bad omen. For Professor Abner Shimony, who had visited the Popper seminar in the previous year from the United States, had been a student of Carl Hempel, whom Popper treated as a chief foe, since he had been an associate of Carnap, the arch enemy, and therefore favorable to induction, which Popper considered the original sin. But, although I missed the "Shimony" slip at the time, it became clear to me on my next and last visit to the seminar that Popper had tagged me as an ally of his philosophical adversaries.

That last visit to the seminar started in the same way as the previous one. Someone, perhaps it was Watkins again but I have forgotten, began to present a paper only to be interrupted by Popper, who revised the sentences just read. Thereafter, the process of short reading, interruption, and revision continued. After about an hour had elapsed, the topic touched on was Popper's well-known criterion of demarcation between science and metaphysics. I had been unhappy with the criterion in question and decided to raise my doubts about it by a query to Popper. Before he could respond, however, the chorus of his disciples led, as I recall, by Agassi, began to shout disparaging comments relating to my question–which startled me. Popper tried to calm his chorus down, with but indifferent success. When the hubbub had quieted finally, he gave me his reply and from the tone of his voice, it was evident he clearly considered mine to be a hostile question.

Since his answer did not seem to me to be adequate, I asked him, politely, a follow-up question on the same point. Now, he raised his voice, and repeated his previous reply in some anger. By this time, I considered it a point of honor not to accept his clearly evasive reply, especially when he was being loudly cheered by his gallery and had breached the normal ethic of scholarly argument by intending to intimidate me. The angrier he became, the calmer I grew, so I quietly rephrased my question, to bring out the issue as clearly as I could. Now, he rose from his chair, red in the face, shouted that he could not go on any longer, and stomped out of the room. This extraordinary display ended my personal contact with Popper forever. I did, however, publish the critique of Popper's criterion which had sparked his outburst, in my *Anatomy of Inquiry*,[7] an account to which neither Popper nor his students ever replied.

In informal discussions with others at L.S.E. after my seminar visits, I found that what to me was a startling experience at the seminar was a phenomenon well known to the locals and the object of various jokes, the standard comment being that Popper's seminar was the exact opposite of everything he preached in *The Open Society and Its Enemies*, to wit: far from being open, the seminar was in fact a closed society; in contrast to his ideal of a unified humanity, the seminar showed a radical division between those who agreed with Popper and all others; in place of his urging the need to treat each person rationally as a source of arguments, he treated his interlocutors as potential enemies; and finally; in lieu of his vaunted value of critical rationality, his seminar brooked no criticism, rational or otherwise.

Popper's insensitivity to the inconsistency between his teaching and his practice was evident at a lecture he gave to a Brandeis University audience many years later, a lecture I attended. The room was packed with academics from different fields and at different levels, from undergraduate students to senior professors, all eager to hear words of wisdom from this visiting celebrity. After his lecture, he took questions from the audience, which was on the whole, deferential to him. Toward the end of the question period, a young undergraduate woman seated toward the back of the lecture hall raised her hand. I do not recall if a mobile microphone was brought to her or if she spoke without such aid. At any rate, she shyly asked the question, which was obviously audible to most of those assembled.

Popper asked her to repeat her question, which she did. He then told her he still had not heard her, and requested that she come forward and stand before him. Obviously flustered and embarrassed, she made her way through the crowded aisles with some difficulty until she reached the front and faced the speaker. He then told her to turn around and face the audience. Finally, he asked her to repeat her question into the standing microphone nearby. When she finally, with some hesitation, managed to get her question out, Popper said to her, in some such words (I paraphrase) "Oh, is that what you're asking? I thought your question was quite different. Here's what you should have asked". Then, he proceeded to state the question he said she should have asked, and calmly proceeded to answer it. Meantime, the student had to remain standing in front of the large audience until the whole process was over, when she finally managed to return to her seat.

Popper's humiliating of the young undergraduate student was in marked contrast to the extreme deference he was capable of showing to those he admired as especially distinguished equals—a contrast well-attested in the case of bullies of all sorts. I witnessed such deference at first hand in a small meeting of the senior philosophy faculty at Harvard around the time of Popper's Brandeis visit.

He had been invited to give a talk at this meeting and, as I recall, the topic he chose was his notion of verisimilitude. The meeting took place in the comfortable Bechtel room of Emerson Hall at Harvard, and the audience comprised only senior Harvard philosophers seated in a small circle facing the speaker. Popper entered and, after a perfunctory nod to the audience, turned his gaze exclusively toward Quine and, greeting him with an ingratiating smile, said he knew Quine probably had not agreed with him in the past but that, after the talk he was about to give, he fully expected that Quine would change his mind.

Thereafter, throughout Popper's whole presentation, his gaze never wavered, focussed altogether on Quine, with occasional asides also directed wholly toward him. When he concluded his talk, he paused portentously and asked of Quine, "Well, have I convinced you?", to which Quine shook his head slowly in the negative.

Apparently disappointed, Popper reluctantly opened the floor for general discussion, in which the main exchange centered on a technical question raised by the Harvard logician Hao Wang. Popper replied to Wang in a dismissive manner whereupon Wang, typically the mildest of men, was so put off that he sprang out of his seat and covered the blackboard with an impressive array of logical formulas in support of his question. Clearly not much interested, Popper continued to dismiss the point, it being clear to the rest of us that the only person in the room he considered trying to convince was indeed Quine, who alone merited his full attention and deference.

Truth to tell, I am sorry I ever met Popper in person. I came to his seminar as an admirer of his works but found myself repelled by his teaching practice and the personality it projected. Such divergences between word and deed are a well-known feature of human nature and are reflected even in the everyday folklore of child rearing: We tell the child "Do as I say, not as I do", when his sharp eyes detect such divergence in our dealings with him.

Nor is it in every case hypocrisy that is at issue. For hypocrisy involves feigning a show of virtue to which one does not in fact subscribe. Even without feigning, people are often simply unaware that their practice does not exemplify norms which they espouse, or exemplifies norms they profess to abhor. What we express is, for whatever reason (including ignorance, inadvertence, unconscious motivation) often at odds with what we exemplify, and children are especially sensitive to the divergence between the two. As they mature, they need to learn that the charge of hypocrisy is not always to the point and that a more forgiving tolerance for human complexity is often the more adequate response. This tolerance is often called for whenever we encounter discrepancies of the sort we are discussing, which are rife in the history of the arts and the sciences as well as in more practical human endeavors.

The tolerance in question has a cost. It requires that we separate word and deed and assess them separately, so that we retain whatever is valuable in the one without losing it by assimilation to the other. In the case of Popper, it requires, for example, that I abstract the value I find in his writing from the poor showing I find evidenced in his practice, neither idealizing the latter because of his intellectual stature, nor denigrating the former because of the personal traits of its author.

A more pleasant encounter I had in London was with R. B. Braithwaite, the Cambridge philosopher of science whose work *Scientific Explanation*[8] I had read before I came. At a London meeting whose sponsorship now escapes me, Braithwaite gave a lecture which I attended. As I recall, he touched on the topic of postulational economy, which provoked a question from me having to do with the Quine and Goodman paper, "Elimination of Extralogical Postulates", that had appeared some years earlier in the *Journal of Symbolic Logic*.[9] As I asked my question, the American accent in which it was couched caused everyone's head to turn in my direction. With the experience of Popper's seminar behind me, I did not know what to expect. Braithwaite, however, received my question graciously, acknowledged that he had not known of the paper I had mentioned and asked me to send him the reference. Furthermore, when we had occasion to meet after his lecture, he invited me to visit him in Cambridge.

My wife, my son Sam, aged 7, and daughter Laurie, aged 4 and I accordingly met Braithwaite in his rooms in Cambridge and we also at the same time were introduced to his wife, Margaret Masterman, who joined him for the occasion. Miss Masterman, known for her interest in Chinese and in the history of science, was a lively conversationalist and she told me at the outset that she disagreed with her husband's and my approach to the philosophy of science. After he had served us sherry, he and Miss Masterman invited us all to have lunch with them in a nearby restaurant where our pleasant conversation continued. However, she had prepared a surprise for our children; she presented them each with a toy pistol, warmly received by them and treasured long after that Cambridge visit. Moreover, in mischievously playful spirit, Miss Masterman made several small bread pellets between lunch courses, and launched them across the table at Sam and Laurie, who were delighted to reciprocate in this game in the august environs of Cambridge University.

I visited a class lecture by Braithwaite at Cambridge. I had arrived at the lecture room early and seated myself well before he entered, clad in his academic robe, and clutching a shopping bag from which he retrieved his notes. His lecture, as I recall, concerned aspects of game theory, in which he was at the time intensely interested. In the following few years, I studied his book *Scientific Explanation*, especially for his views of teleological explanation with which I was then concerned. I published a paper including my critical account of these views and incorporated the substance of this account in my *Anatomy of Inquiry*. In 1964, I was a U.S. delegate to the International Congress of Logic, Methodology, and Philosophy of Science, which met that year in Jerusalem. I was delighted to be able to present a copy of my book on that occasion to Braithwaite, and was pleased to see Miss Masterman again, who reminded me how much she disagreed with her husband and myself on our non-historical approach to philosophy of science. It is a curious point that, despite such disagreement between us, which I acknowledged at the time, both she and I offered criticisms of Kuhn's historically oriented *Structure of Scientific Revolutions*,[10] I in my *Science and Subjectivity*[11] and she in her essay in Lakatos and Musgrave, eds. *Criticism and the Growth of Knowledge*.[12]

Shortly before I arrived in London, I had picked up an issue of *The Listener*, which had reprinted the text of three BBC broadcasts that interested me. The title of the three was "Authority, Responsibility and Education", and they were described as the work of R. S. Peters, a philosopher at Birkbeck College, London.[13] Each of these offered an acute discussion of its title theme in a clear and graceful style. Each seemed a model of the sort of philosophical analysis that I had been trying to promote in the field of education both in my anthology, *Philosophy and Education*,[14] and in my *Language of Education*,[15] on which I was still then at work.

As it happened, Peters and I met for the first time at a meeting of the Aristotelian Society in the Fall of 1958. He was a member of the Birkbeck College faculty headed by C. A. Mace, professor of philosophy and psychology. Peters had written a book on Hobbes,[16] and also a philosophical monograph on *The Concept of Motivation*.[17] He had also, at the suggestion of Mace, reduced to one volume and brought up to date Brett's three volume history of psychology in its relation to philosophy, later published as a paperback by the M.I.T. press.[18] When I met him, I had read only his book on motivation and the text of his BBC lectures, which later also appeared as a book, with the same title.

He and I chatted a bit at the Aristotelian Society meeting and introduced one another. I told him that I had recently read his BBC lectures and that I was glad he was writing as a philosopher of education. At this, he bristled and I could see him recoil at the title I had ascribed to him. He explained firmly that he had worked on Hobbes' political philosophy and on philosophical psychology, but not on philosophy of education. Nevertheless, I told him, by composing his BBC lectures he was *de facto* a philosopher of education.

Years before, I had read a quip by Roy Wood Sellars that there are two ways to make someone religious; one way is by conversion, the other is by definition. Seeking an ally in my project of promoting an analytic philosophy of education, I was trying to bring Peters into the fold, if not by conversion, then by definition. He was certainly not convinced by my effort, but he was exceedingly cordial, and invited me to give a paper at his Birkbeck seminar, and to meet his philosophical colleagues in the Birkbeck common room. I did in fact present a paper to his seminar and had good meetings with his colleagues. In my several contacts with Peters, I was impressed by his wide range of interests, his straightforward analytical style and his evident good sense on topics of common concern bearing on philosophy and on education.

A short while after my return from London to Boston, I had a call from Dean Keppel. He told me that a Mr. Chadbourne Gilpatrick of the Rockefeller Foundation was in town and wanted to meet with me that evening. It turned out that what Gilpatrick wanted was to talk with me about the state of the humanities in general and the prospects of philosophy of education in particular. I knew that his Foundation had been very much involved in promoting improvements in the humanities and extending their influence in the professions. It was the Rockefeller Foundation that had, as I was well aware, initially made it possible for Harvard to appoint Bernard Bailyn and me so as to strengthen history and philosophy at the Education School.

After we had spent a few hours getting acquainted and discussing the situation at Harvard, he asked me what I thought the Foundation could do to help increase the influence of philosophy in the work of the School of Education. I described Peters and his work in London and expressed my strong opinion that he would exercise a salutary influence on the field of education by his teaching and writing at the School. Accordingly, I suggested that the Foundation provide a grant to the School to enable Peters to come for a period as a Visiting Professor. After a few weeks, I heard that the Foundation had indeed accepted my suggestion. The following year, Peters arrived to begin his term as Visiting Professor.

As I recall, he came a short time before the Spring term began. Never having been outside the U.K. before, he was totally unaware of the severity of our typical winters in New England. The terms "Fall term" and "Spring term" are in fact totally misleading. "Fall term" sounds gloomy, dark and cold, whereas in fact it is typically pleasant, moderate in temperature, and often sunny. "Spring term" sounds lovely and bright, but it is typically exceedingly cold, often stormy, with a good deal of snow and ice. When Peters arrived for the first time in my office, having just come from the airport, I saw what he was wearing and my heart fell. Hatless and gloveless, he wore a light overcoat and ordinary street shoes. "Richard", I said, "you'd better equip yourself properly for the coming season." I advised him to get a warm winter hat, gloves, a scarf and boots or overshoes. He went out forthwith and returned in a few hours transformed. As matters turned out, the ensuing winter was unusually severe, but he was now well fitted out for survival.

I had earlier corresponded with him about his prospective teaching, my suggestion being that he teach my usual Introduction to Philosophy of Education lecture course in his own way. My course had been taking as its theme epistemology and education; he decided to teach on ethics and education. He was wonderfully successful as a teacher of our graduate students, and his lectures eventually issued in his pioneering book, *Ethics and Education*,[19] appearing in the same series as the book based on my own lectures, *Conditions of Knowledge*.[20] He was very active during his stay; he attended meetings and lectures, he was available to students and he participated in my epistemology seminar as well, which that year dealt with Ayer's *Foundations of Empirical Knowledge*.[21]

He and I shared an emphasis on rational approaches to philosophical issues and the primacy of reasons in teaching and educational development. We were allies in representing this emphasis in the philosophy of education in our own ways. In occasional debates with students, he remained always steady, always calm, always respectful but always insistent that his interlocutors produce reasons for their claims. In this he was a model of the approach he advocated, and he won students over by his unfailing calm pressure to get at the rationale of positions being put forward. Occasionally, there were funny consequences of this attitude. I remember a student phoning him, before a mid-year examination was scheduled in his class, to say that he could not take the examination because he had broken his arm that morning. "Which arm do you write with?" was Peter's immediate response, before eventually excusing the student from the examination.

Over the course of his teaching, he came to develop a genuine liking for American students. When I asked him to compare them to the English students he was used to teaching, he said he found the American students less well trained in philosophy and less polished in speech and writing. On the other hand, he found them more curious, less cautious, more willing to venture their opinions and take the risks of critical discussion. This opinion is similar to that of some others with whom I have had the opportunity to discuss the question. Once I heard my Harvard colleague Morton White describing an Oxford seminar he had observed, conducted by Gilbert Ryle. Ryle would raise some initial question, said White, and then invite discussion. Rather than this invitation's producing a rush of replies, all the participants, as well as Ryle, lit their pipes, and proceeded to puff away in utter silence. Eventually, the silence becoming increasingly unbearable, some American student or visitor would not be able to contain himself any longer and would say something. The ice having been broken, to everyone's relief, the hubbub of discussion began.

During Peters' stay, we had arranged for him to have a room in the Harvard Faculty Club, which had a reading room and a dining room, and provided a pleasant environment in which he could work and meet other scholars. He was quite happy with the arrangement, and did a good deal of writing in his room, preparing his lectures and working on manuscripts. But as the late Spring and the warmer weather arrived, he confided to me one day that he had a complaint, namely, he was getting fat. He said he was eating too much and had not had a chance to play golf. Questioning him, I learned he had been taking all his meals, breakfast, lunch, and dinner every day at the Faculty Club since he thought that was part of our arrangement for him. Appalled, I alerted him to the fact he was not obliged to eat at the Club, but could walk a block or two toward Harvard Square where he would find eating places galore to suit any appetite he might have. As to golf, I told Dean Keppel that our visitor, an avid golfer, needed access to a course, since he feared he was getting fat. Keppel, with his usual efficiency, set to work to solve the problem and managed to find a golf course not too far, to which Peters would be admitted. With the food and the golf problems resolved, our visitor was delighted to find himself feeling fitter and getting back to his usual weight.

Shortly before the end of term, my wife and I had a small farewell party for him at our apartment, and invited many of his students to say goodbye. It was an exceedingly pleasant occasion, and also a sentimental one. The students were obviously fond of him, and the usually stolid Peters was clearly moved by the affection he had aroused as well as by his own feelings for them.

Some time after he returned to London, the chair in the philosophy of education at the University of London's Institute of Education fell vacant, with the retirement of Professor Louis Arnaud Reid. When asked, I was only too glad to write to the Institute recommending Peters for the chair. He was, in fact, appointed to that chair and I was elated that Peters, who had on our first acquaintance denied being tagged as a philosopher of education, was now going to occupy the most prestigious chair in that subject in the British Commonwealth. I also looked forward to the collaboration he and I would share in promoting the philosophy of education on both sides of the Atlantic. He in fact proceeded to invigorate philosophical training in the network of teachers colleges in the U.K. and his influence extended to the Commonwealth countries as well. By his unusual vigor in teaching and his many writings, he raised the status of philosophy of education immeasurably. We remained fast friends and allies ever after.

CHAPTER 7

SOME HARVARD PHILOSOPHERS

Before I first began teaching at Harvard, I had of course known of several eminent members of the faculty, and when I came to Cambridge I resolved to meet them or at least audit their lectures. Among these, I had read and particularly admired Professor Gordon W. Allport, author of *Personality: A Psychological Interpretation*,[1] and Professor Robert H. Pfeiffer, author of *Introduction to the Old Testament*.[2]

Allport had been the editor of the *Journal of Abnormal and Social Psychology*, and I had had an interesting contact with him when I submitted a short paper to his journal in my student days. Not having had word for a long time that the paper had been either accepted or rejected, I assumed the worst. Then, long afterward, I got a communication from Allport that explained the delay. Apparently, he had gotten another short paper on the same subject at about the same time mine was submitted and he faced a problem. He did not think the topic merited the publication of both papers, nor did he want to accept the one but not the other. Finally, he decided to ask both authors if we would be willing to combine the two papers in one and submit the enlarged version as a joint effort. This struck us as a wise judgment, the opposite, one might say, of King Solomon's famous suggestion to divide the baby, thus managing to save it in the end; here Allport suggested joining the papers, thus saving them both. So, the paper eventually appeared under the joint authorship of Cornelius L. Golightly, whom I never met, before or since, and myself.[3]

I had studied a good deal of Pfeiffer's monumental book on the Old Testament in connection with my biblical studies with the eminent scholar H. L. Ginsberg at the Jewish Theological Seminary, and I looked forward to attending Pfeiffer's lectures, as well as those of Allport. But then, the work of teaching began, along with student advising, committee responsibilities, grading of term papers and examinations etc. and I discovered several years later, to my great disappointment, that many of those Harvard professors I had resolved to meet, including Allport and Pfeiffer, were now deceased.

An eminent Harvard logician, Henry Sheffer, had also been known to me earlier, as the inventor of the "Sheffer stroke", a logical connective which enabled the axioms of the propositional calculus to be reduced to one, thus effecting a significant economy. This achievement had been acknowledged, and hailed as an important result, by Whitehead and Russell, in the second edition of their *Principia Mathematica*.[4] I found, when I arrived at Harvard, that Sheffer was somewhat eccentric and certainly reclusive. When I discovered that one of my undergraduate students, Ronald Dworkin, was one of the elect few who had been admitted to a seminar that Sheffer was then teaching,

I asked Dworkin what topics Sheffer was treating in the seminar. "I can't tell you," said Dworkin. "Why not?", I asked. The reply was that the seminar students had all been required by Sheffer to take an oath beforehand that they would not reveal to any outsider what had gone on in the seminar.

Of course, I realized it was a lost cause for me to attempt access to his classes. Now it was the custom at Harvard, when a professor gave his last class, that senior members of his department would attend in order to do him honor and bid him farewell on this special occasion. Since I was a junior faculty member at the time, I could not join Sheffer's senior colleagues on this mission, but I heard later on that they had, following the custom, entered his classroom before he arrived for his last class, and seated themselves quietly in the rear. Then Sheffer came in, looked around and saw his hand-picked students at the front, but a row of unfamiliar faces at the back which he did not recognize, being extremely nearsighted. "Who are those auditors?" he shouted. "Out, out!" he continued, while his senior colleagues quietly shuffled out, never, of course, revealing their identities to their valued colleague.

Sheffer was one of the genuine faculty eccentrics during my early years at Harvard. Another was the eminent scholar of Christianity Arthur Darby Nock. I never met him personally but would occasionally see him in Widener Library where he cut a striking figure, a tall man, dressed in black carrying a furled umbrella, and with trousers so short that they did not reach the tops of his shoes, leaving about two inches of white socks visible at his ankles. Oblivious to the attention he was arousing, he spoke to the librarian in a loud baritone, clearly audible well beyond the checkout desk where he was transacting his business. In leaving the desk and making his way through the tables of the reading room carrying his books, he would use his umbrella in the manner of a hockey stick. Seeing any student with feet on the table while reclining in his chair studying, he would knock the feet off without a word, startling its owner as he continued on his way.

Many anecdotes circulated about Nock. An apocryphal rumor had it that he sat at the desk in his study in the nude, poring over his scholarly tomes. My favorite anecdote, fanciful though it probably is, concerns a cleaning woman who entered his room in Eliot House and found him seated naked on the floor, meditating. "Jesus Christ!" she cried, to which he replied, "No Madam, only His obedient servant, Arthur Darby Nock".

Although I never succeeded in meeting Sheffer, he and I did, however, have a curious pseudo-encounter. I have already mentioned my attendance as a delegate to the International Congress of Logic, Methodology and Philosophy of Science which met in Jerusalem in 1964. The evening before the first session, there was a reception for the delegates, which I attended. As I entered the room, where people were having drinks and chatting in small groups, I sensed something odd in the atmosphere around me, which I could not quite place. Then, I saw some people staring at me, with peculiar expressions on their faces. Engaging some of them in conversation, I soon found out what lay behind their reactions. "We thought you were dead," said one of them.

And I later conjectured that the recent news of Sheffer's death had mistakenly been transferred to me, because of the similarity of our names. I never crossed paths with him but did, in this way, have some sort of contact with his ghost.

The Congress President that year was Alfred Tarski, and my fellow U.S. delegates included Max Black and Richard Montague. Speakers on the program included Alonzo Church, Michael Polanyi and Donald Davidson as well as other eminent philosophers. The delegates had been notified that an important motion was to be presented at the forthcoming plenary session to take place a day or two later. The Congress was so organized as to welcome those concerned with the natural sciences as distinct from the social sciences, dealing with the logic, methodology and philosophy of the former sciences alone. The motion to be put forward for discussion and a vote was that the social sciences also be included in the framework of the Congress.

Here, an unexpected thing happened. We U.S. delegates, discussing the matter amongst ourselves, were all enthusiastically in favor of the motion. But then, to our surprise, we were beseeched by the "iron curtain" delegates from Eastern Europe, by all means to vote against the motion. Why? As they explained it, if the motion were to pass, then delegations from the Soviet bloc would no longer be restricted to eminent scholars in mathematics and the physical sciences, but would also include so-called social scientists, who, in fact would be secret police masquerading as scholars, and whose job it would be to monitor the conduct of the real scholars and keep them in line. We were convinced, voted accordingly, and the motion failed.

One of the term papers I had submitted as a graduate student in Goodman's seminar at Pennsylvania was, at his suggestion, sent to the editors of *The Journal of Philosophy*, who accepted it for publication.[5] When it appeared, I gave him a reprint. Some weeks later, I received a gracious handwritten letter from Professor Henry Aiken of the Harvard philosophy department, to whom I imagine Goodman had given the reprint and with whom he had spoken about me on a recent visit of his to Cambridge. Aiken's letter mentioned his conversation with Goodman and went on to make complimentary remarks about my paper, which opposed the dualism of mental and physical terms–a position Aiken found congenial. The letter concluded by inviting me to meet with him on any future visit I might make to Cambridge.

At that time, I would have thought the idea of such a meeting totally beyond the realm of possibility. Yet within seven or eight months the impossible happened. I was appointed as an instructor at Harvard and opted to take ethics as the topic of my first graduate seminar. I was advised by Goodman to telephone Aiken in advance and see how he felt about this, since he was the ethicist of the department. Aiken was extremely cordial on the phone and encouraged me by all means to devote my seminar to ethics.

I was exceedingly nervous about this seminar and started experiencing stomach pains a few days before the first class meeting. I thought I'd better consult a doctor since I hadn't had a physical in years, but did not know any doctors in Boston. Seeing the shingle of an M.D. at the end of my street I decided to consult him. It turned out the doctor seemed about 90 years old and, not having seen a patient for a very long time, was eager to keep me in his examining room for as long as possible. He told me about his medical education at Harvard and showed me his stack of recent medical journals which evidenced his keeping up to date with the latest advances. Then came the physical exam which went on for well over an hour and a half, the most thorough checkup I had ever had, during which we talked about medicine, philosophy, Harvard, and the state of the world. I was gratified in the end to learn that nothing at all was wrong with me medically; the only advice the doctor had for me was to drink more water. Thus, somewhat more relaxed, I thanked my genial physician and left his office walking with a lighter step, my nervous pains having mysteriously vanished.

I was nevertheless pretty tense as I entered my seminar room, the seats around the table already occupied, and was greeted by a good looking man of about 40, with sandy hair and a warm smile, who turned out to be Henry Aiken. From the somewhat old-fashioned penmanship of his letter and his eminent position as Harvard's ethicist, I had formed an image of Aiken as a solemn, courtly scholar of the older generation, and it took me a while to readjust my thoughts. Aiken was young in spirit as well as age, lively in demeanor, and vigorous in argument. He took part actively in the ensuing seminar discussion along with the graduate students and myself, and the spirit of the proceedings turned out so engaging that I lost all my fears, impressed though I was with the caliber of the participants.

That first meeting with Aiken was the beginning of a friendship that lasted for decades, until his untimely death. Aiken's main field was ethics, but he also taught aesthetics and the history of philosophy. He wrote on Hume's moral and political philosophy, on the thinkers of the nineteenth century in his *The Age of Ideology*,[6] and on the American pragmatists, among other works; he was also the author of a book, *Reason and Conduct*,[7] and of a large number of papers on problems of analytical ethics. His scholarship was not dry-as-dust, even though it very often dealt with technical issues. He was always concerned with the larger social, moral, and religious import of the thinkers he treated.

Aiken was deeply engaged with music and poetry as well, which gave him access to deep layers of feeling. He comprehended human frailty, was sympathetic to the weaknesses of others and painfully aware of his own vulnerability. All of these aspects of his life gave him the basis for a religious sensibility that was serious and abiding, even if distinct from the conventional forms of organized religion.

Aiken was a dedicated teacher, who enjoyed teaching and was warmly appreciated by his students. One of the architects of the Harvard General Education course, Ideas of Man and the World in Western Civilization, he helped by means of this course to exhibit philosophy as a broad and humane though analytical enterprise rather than a specialism; in this he followed the example of his mentor Ralph Barton Perry and the pragmatists, in particular William James, whom he much admired.

One of his evident characteristics was his liveliness as shown in his sense of fun, his ironic wit, his sociability, and his sensitivity to the arts and literature. It is an interesting fact that, despite his affiliation with the analytic movement in ethics, he collaborated with the existentialist William Barrett in producing an anthology of modern philosophy, *Philosophy in the Twentieth Century*,[8] at a time when analytic philosophers looked with disdain on existentialism. Though he was eager to understand and appreciate philosophical perspectives other than his own, he staunchly defended analytical methods in philosophy. To a mathematics professor who once complained to him that current philosophy was too analytical and not concerned with the grand old metaphysical problems of the past he replied by complaining that current mathematics was no longer concerned with such grand old problems as trisecting the angle and squaring the circle.

Aiken enjoyed unusual characters and hated pomposity. He did not much appreciate the stiffness and austerity of C. I. Lewis, the senior member of his department, but had a genuine affection for the eccentric Sheffer, whom he often enjoyed quoting as having addressed Raphael Demos, the professor of Greek philosophy, at lunch table in the Faculty Club, by saying, "How do, So-crate-eez, please pass the poe-tay-teez." He also quoted Sheffer's quip directed at their colleague Quine: "Quinine logic, a bitter pill."

Aiken had a rather poor opinion of the theologian Paul Tillich, whose obscurity, he thought, was exceeded only by his self-importance. For a time Tillich had a part-time affiliation with the philosophy department in addition to his teaching at the Harvard Divinity School. I remember one occasion at which, in a small seminar of the department, Tillich was invited to present a talk on his views. Aiken, in the ensuing discussion period, launched into a vigorous critique of Tillich's presentation. Here, however, Tillich got the better of him, by using a rare but effective strategy, namely, agreeing with every criticism Aiken made, no matter how damaging. The gist of the exchange (I have forgotten the actual words) went something like this: *Aiken*: But, Professor Tillich, what you have said is self-contradictory. *Tillich*: Professor Aiken, you are absolutely right! This only shows the complexity of the problem. Aiken after a while had to throw in the towel, finding no flaw in Tillich's arguments so damning that Tillich did not cheerfully acknowledge it, with a smile of victory, to boot.

A somewhat related encounter involved an attempt by Aiken and a confederate to show up a younger theologian of the same school of thought as Tillich. This theologian often laced his orations with arcane references to obscure historical figures taken as authorities. Aiken decided to get even. He and his confederate confronted the theologian and, standing on either side of him in a corner of the Faculty Club lounge, began a conversation about a mythical 16th century scholar they had invented for the purpose, a so-called Dietrich von Leipzig. They conducted a conversation with each other, across the theologian's head, about the metaphysical views of the nonexistent Dietrich, waxing ever more confident about their spoof as the theologian remained silent. After a while, Aiken or his confederate turned to the theologian and asked for his opinion on the topics they had been discussing. The theologian, far from nonplussed, pitched right in, offering detailed elaborations of the doctrines of the mythical Dietrich, adding commentary and information about his philosophical associates and his rivals. Aiken and his confederate were beaten. They retreated, left with a gnawing doubt as to whether their invented Dietrich had not in fact been a real historical figure.

Aiken and his wife Helen had wonderful parties at their home, with numerous guests, including philosophers, psychologists, literary people, artists, and others, and featuring singing as well as conversation and argument. At one such evening I remember, Tom Lehrer, who was at that time teaching mathematics at the Harvard Education School, accompanied himself at the piano while singing a number of his well-known ironic songs as others joined in. Henry enjoyed cooking, at which he in fact excelled, having indeed at one time offered a course in cooking at a local adult education school. He often prepared the food he offered his guests at these parties as well as at the more intimate dinners he and Helen hosted. In the lively, outgoing, witty, open and intellectual ambience they provided, their guests had a marvelously good time.

Aiken was great company but in one respect, at least, he was exasperating. In the committee work that was the lot of all members of the Harvard faculty, he was often so distracted or absent-minded that he often could not be relied on to perform the necessary chores. I had my taste of this unreliability when he served as member of a committee I was appointed to chair, the committee to govern the Ph.D. in Education degree, a joint degree of Arts and Sciences and Education. The degree program required each candidate to have an approved course of study supervised by one advisor from the Education faculty and another from one of the Arts and Sciences departments. The committee met regularly to decide on student admissions, monitor student progress, as well as act as liaison between students, supervisors and departments. As a participant in these meetings, Aiken was responsible, sound, and effective. But if a committee chore had to be accomplished outside of these meetings, he was virtually impossible to deal with.

One graduate student in the program had Aiken and myself as his advisors and he was making excellent progress, working with the both of us for a number of years. His thesis had in fact been approved and he had passed his thesis defense. As commencement approached, various forms had to be completed and signed by me as chair of the committee. But one essential document could not thus be handled and that was the page carrying both advisors' signatures certifying their approval of the thesis; that page was eventually to be inserted at the beginning of the bound copy of the thesis destined for the relevant library shelf. Now this document became available for signature only after the regular schedule of committee meetings had been completed, so getting the required signatures had to rely on the U.S. postal service.

When, as chair, I received the document, I therefore signed it at once and sent it by mail to Aiken, who at that time lived in Lexington, as I recall. Emphasizing that the time was short, my covering note asked him to sign and return the document immediately because it had to be transmitted to the printer. No reply came from Aiken the first week. I telephoned him to remind him to sign and return the document, which he assured me he would do forthwith. No reply came after four more days passed, and I became panicky, phoning again and receiving another assurance. Another four days passed and I decided drastic steps were required.

I phoned my student and asked him to meet me next morning at 11:00 in my office. When he arrived, I took him with me to my garage and, without telling him what was up, invited him to get into my car. Then I explained that we were going to drive out to Aiken's house in Lexington. I had meantime taken the precaution of getting a duplicate page and signing it in advance. Without calling ahead, we arrived at Aiken's front door and knocked. He opened the door, obviously surprised to see us, and invited us in. "I'm just about to have lunch," he said, "of course you'll join me!" "Of course, we'll be glad to join you, Henry," I replied, "but first you have to do something for me", and I took out the duplicate page I had prepared, handed it to him with a pen and said, "Now, sign". After he had signed, we all sat down at his table and partook of a delicious lunch he had prepared, my student in an obviously better frame of mind knowing that he would now indeed graduate on time.

I have on and off wondered whether Aiken's irresponsibility in such matters was natural or deliberate. I am inclined to think it was natural in good measure, part of an automatically short attention span attaching to what he considered bureaucratic detail. But it cannot be gainsaid that such irresponsibility carried a welcome reward, namely, his not being asked to serve as chair of any university committee or to take on any comparable office. I have since occasionally advised my exceedingly conscientious students to behave irresponsibly in committee matters after their first university appointments, so as not to be swallowed up in administrative assignments. But my advice, given in half jest, has never been taken. Conscientious academics cannot deliberately act irresponsibly, and many do in fact get burdened with bureaucratic duties beyond their fair share.

After teaching for decades at Harvard, Aiken resigned his post in 1965 in order to accept a professorship at Brandeis University, where he taught for the rest of his academic career. One product of this move was a fascinating book which provides a brilliant discussion of university problems and gives an account of his life at three educational institutions, Reed College, where he did his undergraduate studies, Harvard University, where he taught for so long, and Brandeis University, to which he had lately moved. This book, entitled *The Predicament of the University*,[9] recounts his experiences at these institutions, to each of which he was deeply attached, and he writes with special affection of the General Education course, earlier mentioned, in which he played a key role. An intriguing comparison he occasionally made, in conversation, juxtaposed Harvard and Brandeis with respect to philosophy, suggesting the startling conclusion that Brandeis was more philosophical than Harvard.

Harvard, it is true, he said, has the vastly longer tradition, the incomparably finer library, the more renowned research faculty and the larger and more accomplished student body. But the very strengths of its educational history militate against its philosophical character, Aiken argued. For its institutional structure has been in effect for so long that most procedural questions that might tend to arise have already been settled. The long chains of its precedents have lent it stability, precluding the need to raise questions about fundamentals. By contrast, Brandeis, being a young university, needed still to make decisions about its basic as well as its everyday structures and procedures. As a result, everything was open to questioning and reconsideration. Like the old joke about the psychoanalyst meeting a friend who says "Hello", everybody at Brandeis would tend to respond to such a greeting by wondering, "I wonder what he meant by that?" Every contingency that arose thus became an occasion for philosophizing, that is, for concern with meaning and for asking why. Philosophy at Brandeis, said Aiken, was thus always in play, an everyday need rather than an academic subject only.

One of Aiken's capacities was a talent for mimicry. He could imitate accents and locutions, foreign lilts and local dialects, with such accuracy as to produce gales of wicked laughter in hearers acquainted with his subjects. His mimicry was not, at least for the most part, malicious but done good naturedly, in appreciation as well as exaggeration of the forms he was imitating.

I remember, with affection, his reporting an incident that took place on his first trip abroad, on sabbatical. He and Helen had traveled to Spain and stayed in the city of Tarragona. There he was much taken with the culture, the setting and the townspeople. In particular, he enjoyed the sound of their speech and tried to duplicate their intonations but was not able to do so to his satisfaction. One night, in a bar, he struck up a conversation with a local whose pronunciation of "si" fascinated him. He tried to get it right but did not quite succeed, so he asked his companion to repeat it in order again to try duplicating the sound, but failed once more. Realizing that he would need to have an extended tutorial, he offered to pay his friend one peseta for each "si" he repeated. The deal struck, Henry sat in the bar quite a while, paying not only for drinks but also for extended chains of "si" until he finally felt, like Eliza Doolittle, "By God, I've got it."

The two most well known members of the Harvard philosophy department when I joined the faculty were Professors C. I. Lewis and W. V. Quine. C. I. Lewis, known as a conceptual pragmatist, was esteemed for his contributions to the theory of knowledge as well as his promotion of the field of logic at Harvard. His two books, *Mind and the World Order*[10] and *An Analysis of Knowledge and Valuation*[11] were major works in the theory of knowledge and widely studied.

For me he had a special interest since he had been an important teacher of Nelson Goodman's at Harvard. In addition, Goodman had had us read his *Analysis of Knowledge and Valuation* in one of his graduate seminars, and I had been taken by his plain style and penetrating analytic insights. I eagerly anticipated attending his seminars and, when I was first introduced to him at a departmental meeting, asked his permission to sit in. He said no, he would not allow me to attend, did not in general allow auditors. The only occasion I had thereafter to hear him was at a lecture he gave to an open philosophy meeting at Harvard, the last year of his active teaching.

My relations with the eminent logician, W. V. Quine, were quite different. He had welcomed my arrival, having been briefed about me by Goodman and probably helped my candidacy for appointment. When I was readying a manuscript, based on my thesis, for submission to a journal, I made bold to send him my text and ask for his opinion. He was kind enough to respond almost immediately with a critical comment which enabled me to make my point clearer. Some years later, I sent another manuscript to Gilbert Ryle of Oxford, then editor of *Mind*. A long while passed without my hearing either yea or nay from Ryle. Since Quine was that year Eastman professor at Oxford, I presumed upon his good nature to send him a letter and ask him, when he next had the occasion to see Ryle, to inquire about my paper.

Within about a week, an air letter arrived from Quine saying he had asked Ryle about my manuscript. Ryle had apparently lost it, later writing me to say that he couldn't find it "in any of the heaps where it should be", nor could he locate it "in any of the heaps where it shouldn't be." The result, said Quine, was that it was "a case of out of sight out of mind, and a *fortiori* out of *Mind*." He continued by expressing his hope that I had retained a copy. If not, he wrote, "the matter smells of catastrophe". I had, in fact, kept a copy, and the paper was in due course published by another journal. But I have never forgotten Quine's kindness in responding so promptly to a presumptuous request by a new young instructor.

I had another unsatisfactory contact with Ryle when, undaunted by his unapologetic loss of my paper, I was brash enough to send him another, considerably longer paper for *Mind*. Realizing that it was beyond the length of articles typically published by the journal, I suggested to him that if it didn't fit, perhaps he could consider publishing it in two parts, since I had run across an occasional two-part paper in prior issues. This time Ryle not only did not lose my manuscript but he responded quickly to reject my paper, writing, in some such words as these, "There are only a few people in the world for whom I would publish a paper in two parts, and you are not one of them." This paper was also soon published by an American journal and I appreciated Ryle's responding as quickly as he did, saving me much time.

During the fifties, a number of Harvard students who had visited Oxford brought back personal reports of English ordinary language philosophy and of some of its dominant figures, among them Ryle and Austin. Austin was not yet as well known in the United States as Ryle, whose book, *The Concept of Mind*,[12] had quickly captured a large audience not only because of its bold "logical behaviorism" but also by dint of its ingenious deployment of everyday examples couched in striking vernacular language. "Le style", quipped Austin, "c'est Ryle".

One report by Burton Dreben after his visit to Oxford described Ryle's method of editing *Mind*. According to Dreben, he had a large "heap" of submitted manuscripts on his desk. When a new manuscript was received, it was placed at the bottom of the heap. When the time came for Ryle to assemble the next issue, he would start at the top and select what seemed to him suitable papers. If he approved of a paper, it was set aside for the next issue; if not, it was returned to the bottom once more.

On one occasion when Dreben was in Ryle's room during the editing process, Ryle took a paper off the top, glanced at it and showed it to Dreben saying, "Look at this paper and tell me why it is unsuitable for publication in *Mind*". When Dreben started to read the paper, Ryle interrupted: "I didn't say read the paper, I said look at it." The point was soon made clear to the puzzled Dreben. The paper was unsuitable because it contained footnotes, which Ryle detested. He especially abhorred footnotes that acknowledged the author's colleagues for their help or criticism. "If you want to thank someone, take out an ad in the London Times", he thundered. Hearing these reports, I understood better how Ryle could have managed to lose my manuscript in one of his "heaps".

After Quine's return to Harvard, I did in fact sit in on his lectures, which eventually appeared in the book *Word and Object*[13]. The topics were of intrinsic interest to me, and his teaching was a pleasure to observe. He lectured from index cards, nailing down each point clearly, by general description and illustrative example, only occasionally taking a question from the class, but often lingering a bit after his lecture for further questions. Often, he would begin his class by referring to questions that students had raised the day before, and proceeding to respond to them. His lectures were not completely written out in advance, but they were original and systematic, yet shaped so as to be accessible to the attentive listener. He was not a flashy or dynamic lecturer but his expositions were spiced with intriguing examples and surprising turns of thought, and the whole was clearly still undergoing development, an inspiring feature in itself.

Quine's openness to questions and putative counterexamples was evident. He did not welcome the foolish and the obtuse, but he was quite hospitable to the honest and straightforward objection and the truth seeking inquirer. One possibly apocryphal story concerns a brash undergraduate who some years later managed to ignore Quine's intimidating reputation as a single-minded investigator of the abstruse reaches of logical theory. Overcoming the prevalent reluctance in the student body to violate the sanctity of Quine's study, he boldly knocked on the door one day and was invited to enter. Facing Quine, who had obviously been engrossed in work at his desk, the student said, "Professor Quine, I believe in God, what do you believe in?" To which Quine replied, "I believe in geography", and proceeded to explain his fascination with maps. As a matter of fact, he prided himself on his detailed geographical knowledge and his wide-ranging travels to distant places and strange climes.

Shortly before I left for London in 1958, Noam Chomsky and I had written a short paper criticizing Quine's celebrated criterion of ontological commitment as set forth in his paper "On What There Is" and some other papers included in his book, *From a Logical Point of View*[14]. We arranged a meeting with Quine to state our critical points and get his reactions. His attitude was open, his response in fact exceedingly mild and not at all defensive. We decided then to go ahead, and, having been invited to present a paper to the Aristotelian Society in London during my stay, I offered to present Chomsky's and my joint paper, "What is Said to Be".[15]

The custom at the Aristotelian Society was for the presented papers to be printed and circulated to the membership in advance of their respective sessions. Then, the session chairman would introduce the speaker who, presuming that the paper had already been read by the audience, would make only the briefest presentation whereupon general discussion would ensue. The chairman for my session was Professor Findlay, who launched into a lengthy introduction of my paper, refuting it completely and utterly to his satisfaction, his main point being that the whole problem that Quine had dealt with was inane and his thesis absurd. At that point, he turned the chair over to me to present my paper critical of Quine. Somewhat nonplussed, I had no alternative but, rather than criticizing Quine, to defend him vigorously which I did for the better part of my session, in order to be able to motivate my objections. So in the end, I came not to criticize Quine but to praise him to his English skeptics.

When I returned to Harvard, I related this story to Quine, who got a chuckle out of it. When Henry Aiken later asked Quine about our criticism and queried his not replying to us in print, Quine told him that his criterion, which was the object of our critique, had not been intended by him as a technical thesis but only as a vernacular comment. He must, it seems, have disagreed with our criticism, but never expressed his disagreement either in our early, pre-London meeting, or afterward in print.

A psychiatrist friend of mine once was asked his opinion as to what sort of person makes a good administrator. One who can tolerate a cluttered desk, he replied. At the time I heard that remark, I confirmed it to my satisfaction by applying it to Keppel, the outstanding administrator in my experience, whose desk was always piled high with reports, letters, memoranda, and such, but who always had time to talk to his colleagues about serious matters, undisturbed by the clutter. At the opposite end of the spectrum was Quine, whose desk was always immaculate, clean of everything but the page on which he happened to be working at the moment.

His bookshelves were also models of orderliness, everything in its place, every book accessible by system. The economy thus evident extended also to his habit of not discarding old paper but using the reverse on which to write new thoughts. He once told me his method of dealing with the large numbers of correspondents who sent him queries or criticisms aroused by his many publications. He would reply to them and then, after receiving a follow-up letter from them, would reply a second time but unless the third letter from them clearly showed understanding, would not reply thereafter; thus, he economized time. And, in fact, Quine was not made for administration. Although he was engaged in numerous professional organizations, he did not relish administrative tasks. When he needed to undertake such unavoidable tasks, he executed them efficiently and flawlessly, but they were preferably to be avoided in the first place. The well-known remark he made in one of his philosophical papers about his preference for desert landscapes[16] exemplifies his attitude, both metaphysical and personal.

For me, an admirer of Quine as well as Goodman, who were early collaborators, the contrast they presented in this regard posed a perennial puzzle. For Goodman's office was full of clutter; in the last decades of his life, his study looked as if it had been struck by a cyclone. There were books and papers everywhere, on his desk, his tables, his chairs, in no discernible order, nor was any system evident, to the visitor at least, by which his bookshelves were arranged. In addition there were paintings and pieces of sculpture displayed about the room in contrast to the unrelieved bookishness of the faculty studies in Emerson Hall.

Goodman did, in fact, like Keppel, confirm my psychiatrist friend's remark. He did take on various administrative roles during his career. During his graduate student days, he managed the Walker-Goodman gallery in Boston; later on, he founded the Harvard Dance Center as well as the Arts Management Institute and Project Zero, both at Harvard, and he developed multi-media presentations on drama and other arts, requiring the collaboration of performers fulfilling various roles.

The contrast between Quine and Goodman was, however, puzzling because both, despite their different personal styles had been champions of nominalism, hence of economy. Clearly, they interpreted economy differently. Quine, the admirer of desert landscapes, abjured entities that could not in the end be interpreted as physical, given whatever apparatus physics apparently required to disclose them, even the acknowledgement of classes, anathema to nominalists. Goodman, by contrast to this sort of economy, took his nominalism as refusing to allow classes, though open, so far as nominalism is concerned, to the postulation of any other individual entities in a theoretical system. While Quine assumed the primacy of physics, Goodman argued against such primacy, defending the legitimacy of conflicting systems irreducible to a common base, hence in his terminology representing different worlds. To put the matter in simple although somewhat misleading terms, Quine opted for "ontological" economy; Goodman for "systematic" economy.

The clutter in Goodman's study truly represented his acknowledgement of many worlds; science, literature, the arts and philosophy–including even the emotions, which he defended as instruments of cognition. The sparseness of Quine's study expressed his single-minded devotion to logic and physical sciences, even to the extreme of confessing, in his autobiography, that he was exceedingly moved by poetry, hence avoided it whenever possible. I should avow that my sympathies in these matters are with Goodman rather than Quine; I do not hold to the primacy of physics, since current reduction of all our truths to physics is patently out of the question. For me, reduction is a strategy, not a faith; we ought to reduce wherever we can, but we have to make do at each moment with the unreduced scatter of truths clustered in various ways. I do not, it is true, accept Goodman's terminology of many worlds of our making, but I value his welcoming the multiplicity of orders and systems we need to acknowledge.

The early collaboration of Quine and Goodman, despite their differing predilections in the matter of economy and, more generally, philosophical attitude, shows that agreement is an overrated value in philosophy. Despite their lack of agreement, they produced important joint results, where their separate inclinations converged. More important, their differences issued in different systematic views worth pondering for their own sakes and clear enough to be investigated further, in the common pursuit of philosophical truth to which scholarship ought to be dedicated.

My enormous debt to Goodman did not preclude differences between us. My early work indeed grew out of a close study and extension of some of Goodman's ideas on nominalism as presented in his *Structure of Appearance*[17] and on induction as in his *Fact Fiction and Forecast*;[18] later I profited as well from his theory of symbols, as worked out in his *Languages of Art*.[19] When his fourth major book, *Ways of Worldmaking*,[20] appeared in 1978, I fully anticipated that I would find its approach congenial and instructive. But I was soon taken aback to find that his master idea of "worldmaking" was impossible for me to accept, at least if it were to be taken literally. I could accept the making of world versions but not the making of the things to which they purported to refer.

Thus began a series of exchanges between us that were blunt and uncompromising but did not in the least affect the firm friendship and respect that bound us together. I first voiced my misgivings about worldmaking at an A.P.A. symposium about the book in 1979, in my paper "The Wonderful Worlds of Goodman",[21] and Goodman replied on that occasion and in his *Of Mind and Other Matters*[22]. When my *Inquiries*[23] appeared in 1986, I dedicated the book to Goodman for his eightieth birthday and included in it a reply to his rebuttal. A week or so after I presented the book to Goodman at a small party celebrating his birthday, he phoned to thank me for the book, which he said he had liked "except for one bad patch". He then cautioned me, saying, "Don't think that because you dedicated the book to me, I won't reply to you." "Nelson", I said, "the thought never crossed my mind."

And, true to his word, Goodman did reply in his paper "On Some Worldly Worries.[24] To this I responded in "Worldmaking: Why Worry", in McCormick[25], and Goodman replied in McCormick, op. cit.[26] I responded briefly in a footnote in my *Symbolic Worlds*[27] and with this footnote, I considered the whole series of exchanges between us closed. But I was mistaken. He had been invited to a conference on his work at Nancy in 1997. Before leaving for the conference, Goodman, then 91 years old, asked me if I would write a brief paper outlining my response to his last reply, saying he would take it along, and would either read it at the meeting or at least try to incorporate it or comment on it during the discussion following his talk.

I was extremely reluctant to comply with his request, feeling that we had had our say and that our differences had already shown themselves incapable of resolution. But, as one of the world's great salesmen, he persisted and phoned me several times to press me to comply with his request. He knew and I knew that I would, reluctantly, have to disagree with him once again, but I respected and admired his insistence that my dissenting voice be heard at this important conference. Accordingly I wrote a short paper, "Some Responses to Goodman's Comments in *Starmaking*",[28] the draft of which he took with him.

Some months after the conference which Goodman and I had had no real chance to discuss following his return, he became seriously ill, and he died the following year. Thus ended the long series of exchanges between us outlining our differences on worldmaking. When it was over, I felt I still needed to give my view some independent formulation. The result was my paper "A Plea for Plurealism", given as the Presidential address to the Peirce Society in 1998,[29] in which I tried to combine Peirce's realism with Goodman's pluralism, upholding Goodman's many-worlds doctrine but denying his worldmaking thesis. For me at least, our differences had issued in a positive interpretation which I felt I could accept.

A wonderful Harvard personality I came to know soon after my arrival at Harvard was Philipp Frank. Frank was an eminent scientist, the successor to Einstein's Chair of Theoretical Physics at the German University of Prague, and later a biographer of Einstein. He was appointed to his chair in 1912 and served there for over two and half decades, until he left for the United States in 1938 and found refuge at Harvard in the days of the Hitler terror. Since Harvard had at the time no proper appointment befitting Frank's stature, he was appointed a research associate in physics and philosophy and two years later, he was appointed to a half-time position with tenure as lecturer on physics and mathematics, a post he held until his retirement in 1954, after which he remained active in Cambridge in scientific and philosophical circles for the rest of his life.

I met Frank soon after I arrived in Cambridge in 1952, since I began teaching philosophy of science soon after and this subject was the center of his life and work. The book he published in 1957 was entitled, *Philosophy of Science: The Link Between Science and Philosophy*[30] and the subtitle expressed his credo. In the tradition of philosophically minded scientists, such as Helmholtz, Mach, Poincaré and Einstein, Frank wedded the mastery of primary scientific materials with fundamental reflections on the nature of scientific ideas and their social import. He was an extraordinarily gifted teacher and writer, who could expound difficult subjects for the beginner as well as engage in theoretical analyses at the frontier. For him, philosophy of science was not another technical subject, but a general effort to distill the human meaning of rational inquiry. This attitude engaged his strong interests in the history and sociology of science, and in the latter years of his Harvard work, he involved himself in studies of the national differences that characterize the conduct of scientific research. Such differences are, for him, not ultimate, for rational inquiry has general liberal meaning and universal human appeal; hence his attachment to the positivist movement for unity of science, for which he wrote *Foundations of Physics*, vol. 1, No. 7 of the International Encyclopedia of Unified Science.[31]

Frank's personality was engaging. He enjoyed social as well as intellectual contact with people of different types, with elementary students as well as eminent physicists, and his conversation ranged widely and wittily over current events, political tendencies and scientific as well as philosophical developments. In a habit no doubt nurtured by his Vienna coffee-house experiences, he could often be found from mid-morning until early afternoon in a well known self-service cafeteria on Harvard Square, where he would sit, sipping a cup of coffee while perusing the daily newspaper, welcoming any student or scholar who wanted to join him for a chat.

His sense of humor was always active and was integral with his intelligence, whether applied to weighty or personal matters. I remember one conversation in which he described his experience after an accident, where he lay on the ground waiting for an ambulance. To assure himself that he had suffered no concussion, he set about solving quadratic equations in his head, which restored his peace of mind. The behavior of those around him he found amusing. Until the ambulance arrived, he was told, "Don't move!" But as soon as the paramedics came on the scene, they turned him this way and that, jostling him without mercy.

A story I heard tell of him concerns his serving as a witness at the wedding ceremony of Rudolf Carnap and his wife. The ceremony was conducted in Czech, which Carnap did not understand. Frank therefore acted as translator as well as legal witness. He had to convey the official's questions to Carnap in German, and then translate his answers into Czech for the official to meet the formal requirements of the rite. When the procedure began, Carnap, the meticulous logician and philosopher of language, asked Frank to clarify the meaning of the verbal formulas required. As the procedure continued, Carnap kept interjecting questions as to the logical status of the particular statements he was expected to supply at each juncture. Frank finally interrupted him, saying, in effect, "Do you want to get married or not? If so, just answer and don't ask questions!"

Frank was a champion of the unity of the sciences, by which he meant not only the melding of disparate scientific theories but also the achievement of a large rational outlook that would relate science to moral and political affairs and inform it with historical understanding. The scientific conception, viewed thus broadly, was to be a humane guide to living, and the enterprise of interpreting this conception broadly be viewed as the task of philosophy. Philosophy of science, as in the subtitle of his book, *Philosophy of Science*, was indeed to be understood as "the link between science and philosophy." The urge to promote unification involved the need to stress simple underlying principles to draw separate realms together. Frank thus emphasized the role of simplicity in science itself and theoretical economy as an essential goal, not a mere add-on of scientific method. Without simplicity, he said, there is no science. This particular sentiment was often quoted by Nelson Goodman, who, in his own work, emphasized the inherent connection between the ideal of systematicity and that of simplicity, and devoted a good deal of his work to the analysis of simplicity.

Frank took a leading role in the Institute for the Unity of Science, which brought together scientists and philosophers–a form of social unification dear to his heart. Already in 1929, he had arranged a meeting of philosophers of science with natural scientists by means of a joint session of the German Physical Society, meeting in Prague, with the philosophical Ernst Mach Verein, dealing with Frank's suggested topic, "the epistemology of the exact sciences." And in 1959, he was one of the architects of the Boston Colloquium for the Philosophy of Science, dedicated to creating a forum of scientists, philosophers, and all those interested in promoting a broad, humane and rational view of life.

He participated in the meetings of the Colloquium through 1965, the year before his death. It was through my own early involvement with the Colloquium that I had additional opportunities for contact with Frank, and further occasion to appreciate his good sense and his ability to cut through formality and complexity to the heart of an issue.

On a few occasions when my wife and I were invited to his house for a social evening, we met some of the eminent colleagues and friends who were close to Frank, of whom I remember most vividly Percy Bridgman, author of *The Logic of Modern Physics*.[32] With his gentle and humorous personality, Frank and his wife, Hania, made every such occasion memorable. As a result of one such meeting, Mrs. Frank, who had no children of her own, was interested to learn about our young children. She invited my wife to visit her some afternoon with them. At that visit which took place soon after, Mrs. Frank talked easily and naturally with our children and served them cakes and sweets. But the highlight of the visit was Mrs. Frank's playing the piano and dancing for them. Philipp and Hania were a devoted and loving pair, who had experienced many trying as well as happy times together. The book, *Philosophy of Science* bears the inscription, "Dedicated to Hania in remembrance of our going through the Old and the New World".

The year before Frank's death, the Boston Colloquium prepared a Festschrift in his honor.[33] Robert Cohen, who directed the Colloquium, described the Festschrift as containing along with pages of friendship from his students and colleagues, the proceedings of two years of the Boston Colloquium, as though we wanted to show Philipp Frank that his latest grand-child was now walking a bit. Cohen added that a student read large parts of the book to him, and he seemed pleased. As Cohen later told me, when he presented the volume to Frank in the nursing home where he had been living, Frank turned to the photograph of himself at the front, stared at it meditatively for awhile, then turned to Cohen and said, "What a good shave I had then."

One person who made a strong impression on me was not an Education faculty member but the President of Harvard University, James Bryant Conant, whose public life as President of Harvard, as science advisor to the government, then as High Commissioner of Germany is well known, and recounted as well in his autobiographical *My Several Lives*.[34] Conant came to the Harvard presidency from a chemistry professorship and he made fundamental changes in the university's structure and procedures.

He wanted to transform the university from an Eastern establishment into a national institution, and to ensure that the poor as well as the rich would be enabled to study there. He wanted also to ensure that the faculty consist of the ablest scholars who could be found, and to this end initiated the rigorous system of selection of faculty members with the crucial help of external ad hoc committees. He also began the practice of appointing visiting committees for each department and School of the university, who would meet, consult, evaluate, and advise as to the respective status of every such unit. In initiatives of this sort, Conant left his permanent mark on the university.

When he became president, he was advised that he had two kittens to drown, the reference being to the then Engineering School and the School of Education. He decided to eliminate the Engineering School, since M.I.T. already existed and did not need to be duplicated. But he kept the School of Education, out of a deep concern for the health of the nation's educational system and the prospect of its improvement. It was Conant who in 1946 plucked the young Francis Keppel out of a junior position as an undergraduate admissions officer and appointed him Dean of the School of Education, seeing in him the promise of leadership, despite his lacking any graduate degree, academic Education background or training in administration. With Keppel as Dean, he pioneered the Master of Arts in Teaching program, designed to stimulate the choice of teaching careers by the ablest Arts and Science undergraduates, and together they strove to raise academic and research standards in the field of Education.

Conant's practice, as President, was to visit each of Harvard's graduate schools and to chair their faculty meetings on some sort of rotating basis. Thus, having known of him only by reputation, I was surprised and pleased to see him in person when, shortly after I arrived at Harvard, he chaired an Education faculty meeting, seated next to Keppel at the head of the table in front. His manner was mild and straightforward and Keppel and he interacted pleasantly in the course of the meeting and the discussions with the faculty that were involved.

Aside from his administrative duties as President, Conant had insisted on teaching during his tenure. His focus was not on chemistry, but on the history and philosophy of science. He had initiated a series of studies at the Harvard Press, under the title, "Source Books in The History of Science" and inaugurated a Committee on the History of Science, which later achieved the status of a regular department in Arts and Sciences. His course had two teaching assistants who later became well known figures in the field of history and philosophy of science, Thomas S. Kuhn, author of *The Structure of Scientific Revolutions*,[35] and Leonard K. Nash, author of *The Nature of the Natural Sciences*.[36] Conant also wrote important books in the general area, including his *Modern Science and Modern Man*,[37] and *On Understanding Science*.[38]

Conant's interest in teaching extended also to the higher education curriculum. He authorized the faculty work on general education in the undergraduate college in the forties, which issued in the landmark volume that came to be called "the Redbook", i.e. *General Education in a Free Society*.[39] The motivation was to countervail the excessive specialization consequent upon an unrestrained elective system, by requiring a common core of studies deemed to underlie intelligent participation in a democratic society.

When I came to Harvard, Conant was nearing the end of his Presidency, a celebrated educational statesman and advisor to our government as well as a well known scientist and author. You can conceive my surprise when I, a new instructor, received an invitation from him to join him for lunch at the Faculty Club. I had no idea what lay behind this invitation and could not imagine what he wanted to talk about with me.

As it happened, he wanted to get to know me and to talk about philosophy, in particular to discuss what aspects of philosophy were most important to teach at the Education School. Honored by his invitation and enormously impressed by his long record of achievements, I had not known what to expect. But the whole affair was low key and absolutely egalitarian. He asked for my opinions on the philosophy of science and he ventured his own. At one point, he recounted the views which came later to be generally ascribed to Kuhn and here I offered some of the objections I published many years later in my *Science and Subjectivity*.[40]

When this phase of our discussion was over, he asked for my views on the role of philosophy at the Education School and I was able to give him only a sketchy outline of what I hoped to do as I had only just begun and was still feeling my way. I suggested an emphasis on American philosophy since I planned to teach the pragmatists, and special attention to ethics and the theory of knowledge. It was the latter suggestion which clearly aroused his interest, sufficiently so for him to state his main idea, one which surprised me. He thought it of prime importance for aspiring educators to learn Hume's theory of causality. I did not probe his idea too deeply for it struck me as so odd as to leave me without relevant comment. Our lunch was anyhow drawing to a close. We made our cordial goodbyes, and Conant strode out of the Faculty Club, waving to several of the diners still at table. How fortunate, I thought, that Harvard had at its helm this shrewd, intelligent, and humane educator to steer it through times of trouble.

CHAPTER 8

TURBULENCE IN THE 60's

Conant's irenic vision did not survive the troubled sixties, which saw a vigorous national opposition to the Viet Nam war, an energized civil rights initiative, an increased racial, ethnic and gender self-consciousness, triggered initially by the black students' movement, and a prolonged attack on university traditions of teaching and governance, which included not only peaceful demonstrations but also student strikes, sit-ins and various forms of intimidation and violence. Nathan Pusey, Conant's successor as President of Harvard, gained instant notoriety by calling in the police to break up a student demonstration and terminate a prolonged sit-in.

The student disruptions of normal university processes splintered the faculties as well, many professors sympathetic, in one or another way, with the students, others, in one or another way, opposed. The Arts and Sciences faculty at Harvard saw the spontaneous formation of two groups, a "conservative caucus" and a "liberal caucus", the former upholding traditional university practices, the latter seeing merit in some of the changes pressed by activist student groups. The story has been widely rehearsed in a variety of media with various emphases and value judgments attached.

My purpose here is not to add another general account of these troubled times but to relate certain events which affected me, some in the Graduate School of Education, others in Arts and Sciences. The Education faculty was, on the whole, much softer than the Arts and Sciences faculty, much readier to yield to the demands various student groups made with respect to educational processes and university governance. The Education School was much younger and weaker than the Arts and Sciences School, its faculty more varied, less oriented to the traditional disciplines and more predisposed, by progressive educational ideology, to empathize with the students' point of view. Many members of the faculties indeed exaggerated the wisdom and virtue of student youth, tending to react to increasing demands by increasing accommodation, fearful of setting limits and defining boundaries. Such attitudes were indeed more pronounced in Education than in Arts and Sciences; where the idea of a "conservative caucus" in the Education faculty would, at the time, have seemed incredible.

Only Frederick Olafson and I, both philosophers in the Education School, typically upheld traditional practices of teaching and governance at the School, in the face of a rising tide of opposition. I remember one occasion when an Education colleague came to my office to try, by reasoning with me, to persuade me of the error of my ways. He warned me that I was in danger of being left behind by the wave of the future, in effect an old fogy tied to the past. Wouldn't I rather join the ranks of the youth and join their march toward a glorious tomorrow? I politely said no and he left, shaking his head at my foolish obstinacy. From his point of view and those who thought as he did, I was mired in the dead habits of the past. From Olafson's and my point of view, our faculty antagonists were insensitive to the values of the university, and, in their readiness to yield to attack on these values, lacking in both dignity and courage.

The issues that divided Olafson and me from our colleagues were not, in the large sense, political. We did not criticize opposition to the war nor did we oppose civil rights activities, or demonstrations in advocacy of any changes in university practices. We argued against student voting representation on faculty appointment committees, we deplored the loosening of academic requirements and the weakening of standards, for example, the steady decrease in courses requiring examinations or letter grades and the lowering of graduation criteria, and we fought against admission processes and faculty appointment procedures that placed a preponderant emphasis on racial or ethnic identification.

One such issue that came home to me in a personal way related to student admission practices. When I had first joined the Harvard faculty, or soon thereafter, our admissions procedures banned the inclusion of photographs in application documents to prevent bias for or against an applicant on the basis of racial affiliation. When affirmative action policies began to be implemented, the negative attitude toward racially blind admissions was reversed, primarily in an effort to increase the numbers of black applicants admitted.

The new attitude tended to generalize itself, attaching to a variety of ethnic groupings as well as to the sexes, each increasingly construed as having a prima facie claim to representation in the admitted cohort proportional to its numbers in the population. As this tendency continued, members of the faculty were encouraged to think of themselves not only as scholars or researchers but also as ethnic as well as gender representatives charged with promoting the admission of their several designated flocks.

At one meeting of the admissions committee, one of my colleagues proposed that I, being Jewish, should be designated as the official receiver and processor of applications by Jewish applicants. This proposal seemed to me abhorrent on its face, intruding racial or ethnic criteria into the very heart of the university, balkanizing the faculty as well as the student body. Having rejected the proposal in no uncertain terms, I thought the matter was settled. But I was wrong. The author of the proposal wrote to me to convince me of its justice. There followed an exchange of letters which I regret to this day not having saved so that they might be reproduced here. In the end, I won that skirmish as it concerned me personally, but progressive divisions among faculty members and student bodies proceeded apace, with the ancient vocation of scholarship in pursuit of the truth increasingly threatened by the more ancient tribalism of racial and ethnic groupings.

National anti-war sentiment became more and more active as the war continued, with rallies, teach-ins, sit-ins and demonstrations of one or another kind taking place in the universities as the political atmosphere reached fever pitch. The anti-war movement was by no means peculiar to students but was widely shared by faculty members nation-wide. However, universities across the country varied widely in the degree of disruption that took place in their normal academic work. The opposition to U.S. involvement in the war spilled over into a variety of attacks on the university itself, some within the bounds of legitimate criticism and advocacy, others clear infringements on the freedom to learn and to teach, at the heart of academic freedom.

Adding to the ferment at the universities was the fact that various external groups became active on campus, joining forces with congenial student parties pressing for diverse radical causes, most of them to one or another degree Marxist, some Maoist, all authoritarian in character. One of these groups took aim at me personally, along with a couple of my colleagues, in the most bizarre incident of my academic career, an incident I will recount shortly.

I was, at the time, teaching my philosophy of science course in the Department of Philosophy. The course met in Emerson Hall and the windows of my second floor classroom looked out on the Harvard Yard, as did the windows of my office on the floor below. A certain amount of departure from academic civility and routine had by this time become the norm. Thus, the small vestibule in front of my office had been taken over by one of the radical groups which used it to store its pamphlets. When I opened the vestibule door one day in order to enter my office door within, I found this normally private space piled high at one end with such pamphlets, and, with my key in my office lock, was suddenly confronted by a man who, without knocking, entered my vestibule and demanded, "And what have you done today to end the war?" Startled, I did not reply but continued into my office.

That day, a strike had been called by various student groups, which demanded a cessation of all classes and organized a rally in the Yard. From my office window, I could see the crowd beginning to assemble. Many faculty members indeed did call off their classes that day and many did not. I was one of the latter and proceeded upstairs to meet my class.

Of the twenty or so enrolled students in the class, only three or four showed up; the rest, I assume, attended the rally. Since it was a Spring day, my windows were not shut tight, and the amplified oratory and applause of the crowd were audible in my room. Looking out the window, I could see the Yard virtually as full as on a Commencement Day, with the speakers on the steps of Memorial Church at one end and the assembled listeners extending up the steps of the Widener Library at the other. Moreover, I thought I could recognize the main student speaker as one of the star students in my class, wearing a red headband and holding forth in an impressive way, to the cheers of the crowd. Despite the distraction, I held my class and gave the same lecture I would have given if the whole class had attended. The rally speaker in fact did turn out to be the star student I had recognized and he did in fact earn an A in my course despite having missed my lecture.

The incident I earlier promised to recount began one day when I received in my Larsen Hall mailbox a flyer that named me, along with the historian Oscar Handlin and the sociologist Seymour Martin Lipset as the three faculty members who ran the university, and summoned us to a trial to answer their charges of having committed various misdeeds, the details of which I have forgotten. The flyer was signed "National Caucus of Labor Committees", a name that meant little to me at the time. I vaguely recalled that it was they who had distributed an advertisement some weeks earlier attacking my eminent colleague, W. V. Quine, as an "epistemological idiot".

Thinking no more of the flyer, I walked over to Emerson Hall to give my regular class lecture which met there. Pausing in front of the locked bulletin board on the ground floor, I stopped to read the latest announcements. To my astonishment, there, among the official notices of the day, was a copy of my infamous flyer. How it had managed to receive permission to be posted on the university bulletin board was beyond me. Somewhat rattled and more than a little angry, I climbed the staircase to my classroom and made my way through the waiting students to get to the door, which was opening to let the previous class out.

As I approached the door, I noticed a young woman at the door handing out notices to members of my class as they entered the room. When I got to the door, she handed me one as well and addressed me as she did so, saying, "Well, are you going to attend your trial?" Glancing at the notice in my hand, I saw that it was another copy of my "summons", and I replied to her, saying, "No, I don't respond to unsigned letters, but if you want to talk with me in my office, I would be glad to talk with you. See me after my class and we'll arrange a time." She was waiting for me in the corridor after my class and we did in fact set an appointment, for a day or so later.

She arrived at the arranged time, laden with several copies of a tabloid-sized newspaper, named "New Solidarity". I ushered her into my office and offered her a seat facing me at my desk. "What is it that you want?" I asked her. She had apparently decided not to attack me personally at this point. Instead, she launched into an attack on the university as lacking real communists on the faculty. I mentioned Students for a Democratic Society and Progressive Labor sympathizers on the faculty, including one self-proclaimed communist. "Oh, they are to the right of the Republican party", she said. "You don't have a genuine communist."

Thereafter, she proceeded to sing the praises of the head of the National Caucus of Labor Committees, one Lyn Marcus by name, who, she assured me, was the greatest economist in the world and whose book had been refused publication in virtue of some right wing collusion among publishers. She declared that it was Marcus who truly merited a faculty position and that, without him, Harvard could hardly claim to be really democratic.

I asked her whether, if she could appoint genuine communists to the faculty she would also allow those with anti-communist views to be faculty members as well. This simple question, which I had intended merely as a way of drawing her out further, threw her into utter confusion. She hemmed and hawed, hesitated and stammered, and her attempt at a response eventually ground to a complete halt. She was in fact embarrassed. I told her she ought to give my question a good deal of thought.

Appreciating the truce I had offered her, she asked if she could leave me her stack of "New Solidarity" newspapers and asked further if I would read them. I promised I would, and told her I would be glad to meet with her again at her request and that she could phone me any time for another appointment.

I never heard from her again. Moreover, the original flyer was never followed up by any further contacts from the National Caucus of Labor Committees. I do not believe she was a Harvard student, more likely a naïve adherent of the group mesmerized by Marcus, also unaffiliated with Harvard, but attempting to gain members on campus. She struck me as sad and I was convinced that no one before me at the university had ever taken her seriously enough to talk with her. For awhile I toyed with the idea of interesting some publisher in taking on Marcus' book, and so knocking the wind out of his sails.

It was only later that the full dimensions of the incident became clear. "Lyn Marcus" was apparently an acronym for "Lenin,Marx", and the bearer was none other than Lyndon LaRouche, well known extremist cult leader with a band of deluded followers hypnotized by his bizarre doctrines and his authoritarian personality. He had his followers collecting donations in various public places, such as airports, and eventually ran afoul of the law for credit card fraud as well as other offenses, serving prison time.

I wondered at the time how he or his group had managed to pick me out, along with Professors Handlin and Lipset, as one of the managers of the university. This was especially ironic in my own case for, unlike my two colleagues who were well known in the Arts and Sciences faculty, I was primarily active in the Education faculty and had attended Arts and Science faculty meetings only four or five times in all, never participating actively in any way. Furthermore, I had only a nodding acquaintance with Handlin and Lipset. We knew each other but were hardly close, and each of us was primarily engaged in his own scholarly studies. As fellow members of a ruling cabal, we were singularly unfit, individually and collectively, to do the heavy work of running the university. The charge was in fact bizarre.

Pondering the matter, I eventually concluded that LaRouche had picked our three names off the announcement of a then newly formed organization, University Centers for Rational Alternatives (UCRA) founded by Sidney Hook, dedicated to upholding traditional academic freedoms at the university and opposing its politicization. Be that as it may, there was one final ironic touch to the story. Lipset, the very Lipset targeted for trial by the National Caucus, was at the time scheduled to appear for the defense in a pending civil trial of certain of its members.

In the turmoil of the time, one strange academic reversal affected the philosophy department. The challenge I mentioned earlier, "And what have you done today to end the war?" was initially hurled, as a rhetorical accusation, against the ivory tower faculties, including philosophy, which, for an extended period, saw the number of its student concentrators decline. Then, this number began to increase, while social and political science concentrators began to diminish sharply. The explanation seemed to me to be this: In the first phase, philosophy was deemed insufficiently active in anti-war initiatives. In the later phase, philosophy and other ivory tower pursuits gained the good graces of students because, unlike the social and political sciences, they were clearly useless to the military-industrial complex, hence clearly not pro-war. Their very uselessness was an asset, assuring their campus popularity in those difficult times.

The mid-sixties provided me with a pleasant and productive respite on the occasion of a welcome sabbatical. I have recounted the time I spent in London for my first sabbatical in 1958–9. This second sabbatical, in the academic year 1965–6, was spent at Harvard, at the Center for Cognitive Studies, directed by the psychologists Jerome Bruner and George Miller. I was fortunate to be welcomed at the Center and given a study there in William James Hall, for this arrangement removed me from my usual office location and normal contacts and gave me the freedom to put my mind to my own work.

This work centered on the manuscript of my Mead-Swing lectures, "Science and Subjectivity" delivered at Oberlin earlier in 1965, the burden of which was to rebut the recent attacks on scientific objectivity then gathering steam. My time at Oberlin had been exceedingly pleasant and my four lectures, given on separate days, had been very well received. A former student, Ira Steinberg, had recently joined the Oberlin faculty and he was able to brief me about the College and about notable events on campus. In my free time, I recall attending class lectures on modern art on Steinberg's recommendation, which were in fact splendid, and coming to know several of the Oberlin professors, a most intelligent and friendly group.

Over lunch at the faculty club one day, Professor Anderson of the physics department told an anecdote that made an amusing reference to Harvard, where he had started his teaching career. The Harvard students had, for many years, published a confidential guide to courses for the edification of incoming classes. This publication was notorious for its brutally frank anonymous appraisals of the faculty as well as of the courses themselves. Non-tenured faculty members in particular awaited the "Confi-Guide" 's publication each year with trepidation, and quickly turned the pages to see how they had fared. Professor Anderson recounted his nervous dread when, as a new faculty member, he leafed through the Guide to locate his entry, relieved finally at finding the assessment of his teaching which read, in toto, "Anderson is adequate."

The sabbatical project I had planned for my stay at the Cognitive Studies Center was to be a revision and elaboration of the text of my Oberlin lectures and the addition of a new fifth chapter, the whole to be readied for submission to a publisher. The work went well and the resulting book, *Science and Subjectivity*, was published in 1967 by Bobbs-Merrill, with a second edition by Hackett in 1982.[1] Aside from the writing, I was fortunate to be a resident at the Center, where I was able to meet a number of other scholars visiting for the year, most of them psychologists or working in areas relating to psychology. The interest in Chomsky's work in linguistics, with its obvious connections to psychology and philosophy, was widespread at the Center, and the cognitive studies movement within psychology also had special appeal for philosophers. Thus my identity as philosopher assured me a warm welcome in the ambience of the Center.

A weekly seminar arranged by Bruner, where current research by Center residents or visitors was discussed, facilitated scholarly interchange. An additional feature of William James Hall, the psychology building, independently promoted such interchange. This was a smallish dining room on one of the middle floors, the only place to get lunch in this building of many floors, where otherwise the denizens of each were segregrated from all the rest. The uniqueness and smallness of the dining room guaranteed human contact across floors. First, all residents of the Hall had to use the same set of elevators to get to the dining room and, secondly, the smallness of the room virtually ensured that one had to sit with diners from foreign floors and alien specialties and so, despite oneself, learn something about worlds beyond. When I returned to my regular office after my sabbatical, I lobbied for provision of a small lunchroom at the Education School, to accomplish an analogous commingling of the specialties among the faculty. A few years later, such a lunchroom was in fact built in the basement of Longfellow Hall, and it has thrived happily ever since.

I was fortunate in having as a fellow Center resident for the year Marx Wartofsky of the Boston University Philosophy Department, with whom I collaborated in holding a weekly discussion group to which a few others came, some regularly some not. Wartofsky was a learned and an energetic scholar, a brilliant teacher and a genial colleague. His philosophical interests ranged widely, from history of philosophy, with special expertise in Marxist thought, to philosophy of science, from aesthetics to epistemology, from philosophy of history to metaphysics. As one of the directors, along with Robert Cohen, of the Boston Colloquium for The Philosophy of Science, he helped to create a significant forum for philosophical exchange across varied schools and current intellectual trends.

He and I began to meet regularly for lunch at the start of the academic year and, joined occasionally by others, we continued thus until the end. He was at the time, I believe, working on his philosophy of science book, which paralleled my own project dealing with scientific objectivity. We decided, after a while, to choose a text and discuss it systematically, so that our discussions would be continuous and consequential. For this purpose, we fixed on Michael Polanyi's *Personal Knowledge*.[2]

This book had been in the air for some years but had not been taken seriously or discussed closely by analytical philosophers. Neither Marx nor I had read it word for word, but we were intrigued by the prospect of studying it together in our newborn discussion group. We were aware of Adolf Grünbaum's critique of Polanyi as, in effect, an intuitionist and we sympathized with Grünbaum's strictures. We were also suspicious of Polanyi's description of his book as expressing an existentialist philosophy of science, which struck us as something of an oxymoron. On the other hand, we welcomed Polanyi's breadth and his critical examples taken from the history of science, from psychology, from medicine, from language and from social science.

Surely, we thought, the richness of his canvas and the acuteness of his treatments needed to be taken seriously by analytical and linguistic philosophers then working in self-imposed unawareness of broader currents of thought.

Polanyi had visited the Boston Colloquium where Marx had probably met him but I do not recall if he had visited prior to 1965. I had myself, however, met him briefly at Harvard in 1961 or thereabouts and found him to be an intelligent and charming man. As chairman of the General Problems section of The International Congress for Logic, Philosophy and Methodology of Science scheduled to meet in Jerusalem in 1964, I invited him to give the plenary address in my section. At the Congress, I had further opportunity to talk with him and valued his address which dealt with the role of metaphor in science. As a result of these contacts, I had formed a quite favorable view of him as an original intellect and a scholar who had important things to say, and I deplored the isolation in which he had apparently had to work in Oxford, snubbed by the reigning analytic and linguistic coteries. As an aside, let me say how shocked I was to see, many years after the Congress, that the text of his plenary address had been omitted from the Proceedings, presumably because the topic of metaphor had been judged too soft—a scandalous display of intellectual prejudice.

At any rate, Marx and I assigned ourselves and our several other discussion regulars the task of reading Polanyi's first chapter, to be discussed at our next session. Thereafter, we took succeeding sections as our topics and the discussions turned out to be unusually interesting and productive, treating such themes as perception, faith and authority in science from both a philosophical and historical point of view. At one of the early sessions we found that Marx had written, in large block letters, on the blackboard of our meeting room, "Primum mangiare deinde philosophare", i.e. first we eat, then we philosophize; I had myself proclaimed, in an adaptation of Napoleon's dictum, "One cannot philosophize on an empty stomach". Armed with such counsel, we ate our brown bag lunches first, having on meeting days waived our visits to the common dining room and then, properly fortified, we fell to with a will, engaging in energetic philosophical discussion for the next two hours.

In trying to recall the others who occasionally joined us for discussion, my memory is not altogether clear but I believe that some sessions may have had Robert Schwartz, Peter Achinstein, and Michael Martin as visitors. When we eventually completed the reading of *Personal Knowledge*, we continued with discussion of other topics that bore on common interests, ending the year with a session on the analysis of observation. This topic was then under special scrutiny as a result of the idea that scientific observation is theory-laden, hence no firm source of objectivity. To this session, we invited Quine to present his view of observation, and we invited Goodman to participate and comment on Quine's views. As expected, they disagreed at a very fundamental level and provided us with a splendid finale for the year.

CHAPTER 9

CHICAGO AND JERUSALEM COLLEAGUES

I had heard of Ralph Tyler well before I ever met him. He was the legendary founder and director of the Center for Advanced Study in the Behaviorial Sciences at Stanford and, before that, he had been the head of the Division of Social Sciences at the University of Chicago. In the field of Education, he was well known as the author of a major book on curriculum development. But I had had no occasion to meet him personally until the early sixties, as a result of the initiative by the historian Lawrence Cremin and myself to found the National Academy of Education.

This initiative was first suggested by Robert Ulich several years earlier, who thought there should be a learned society of scholars of educational studies bringing together all the specialties and focussing them on common problems. Cremin and I had often talked about this idea and decided to take steps. First, Cremin collected the constitutions of several learned societies, and drafted a possible constitution for an Education Academy, which he and I then discussed. Next, we met with a number of selected scholars to gauge their reactions to the idea; I recall meeting with John Gardner, for example, in this connection. Finally we convened a larger group of established scholars and administrators in Education, to discuss the nature and feasibility of an Education Academy. We met at the Stanford Center, under the chairmanship of Ralph Tyler, whom I encountered for the first time on that occasion. The rest, as the saying goes, is history. The Academy was duly founded in 1965 and has thrived ever since.

Ralph Tyler, at the time, exhibited his signature characteristics as chairman: A relatively short man who smoked long cigars, he was mild mannered, extremely acute, and had a twinkle in his eye. An easygoing chairman, he gave ample opportunity for everyone to have his say. When, occasionally, he would contribute something, his comments were invariably penetrating, sensible, and wise. I admired his performance but did not get to know him personally until a few years later. His former student at Chicago, Seymour Fox, who became the Director of the School of Education at the Hebrew University in 1967, decided to appoint an external committee of advisers, who would periodically visit the School, review its programs, and advise on the value of its various activities. To this committee, Fox appointed Tyler and Joseph Schwab, another teacher of his at Chicago, as well as Ernest Hilgard, the noted psychologist of Stanford University, and myself, later adding Lawrence Cremin as well. This group met, irregularly, with the faculty and the staff of the School, each time for a period of a week to ten days and we gave our opinions on the various academic and research projects the School was sponsoring.

In these advisory meetings, I was again impressed with Tyler's informed good sense on a wide array of issues which he addressed, both theoretically and practically, always in his low key manner. He listened a lot to what others had to say, respectfully asked occasional questions that went to the heart of the issues, and on and off made comments that revealed new sides of the questions being deliberated. His manner was unfailingly mild, his speech quiet and slow. Often it seemed as if he had fallen asleep and had not heard a word of the discussion going on around him, when suddenly, he would come to life and say something totally pertinent and incisively original.

During the several visits of our advisory committee to the School in Jerusalem, I got to know Tyler much better on his informal and personal side. For one thing, he was altogether unflappable, and this aspect came out several times on the various trips our committee took in Israel during our visits. One trip took us from Jerusalem to Jaffa where we had restaurant reservations for dinner that evening. The driver, who had been hired for the occasion, lost his way and we were nowhere near our destination by our appointed dinner time. Since Tyler was the oldest member of our group, the rest of us were increasingly concerned that he was without nourishment. We asked him if he was getting hungry; we knew that we, his younger colleagues, were certainly by this time both hungry and thirsty. No, he said, I'm fine. Finally, about an hour later we arrived at the restaurant and all was well. During the first course, we asked Tyler again how he had managed during the long delay, to which he replied, "I make it a habit never to get hungry until the food is on the table before me."

On another such trip, our committee arranged an afternoon of leisure for a swim at the shore. We all got into our bathing suits and prepared for our first dip, when Tyler stepped on a piece of glass hidden in the sand, and cut an ugly gash in the sole of his foot. We ran to get someone from the first aid station nearby while Tyler sat where the accident had occurred. Soon a muscular young lifeguard in bathing trunks appeared, holding a vial of an evil looking compound. First, he looked at the bleeding sole of Tyler's foot, and pooh-poohed his injury, showing us the scars on his own torso resulting from a battle during the six-day war of 1967. Then, he raised Tyler's foot and poured some of the vial's ominous liquid directly onto Tyler's wound, causing the rest of us to recoil. Tyler, however, despite his initial wince, immediately regained his poise, and insisted that the rest of us take our swim. Indeed, it took a while for us to override his protests and to lead him, alternately hobbling and hopping on one foot, back to the car for the return to our hotel.

In a conversation I had with him during one of our times in Jerusalem, we happened on the topic of disturbed students and I recounted some of my own encounters with such students over the years. Nothing in my experience, however, matched the incident he related, when a student barged into his office, while he was head of the Social Sciences Division of the University of Chicago, brandishing a knife and threatening to kill him. With that, he ended his tale. "Well, what happened?" I asked. "Nothing", he replied. "What did you do?" I fairly shouted. "I talked with him", said Tyler. "And then what?" I asked. "He put down the knife", was the reply, "and we talked some more, until the police arrived."

Tyler had a humorous and mischievous side which also came out during our conversations between committee meetings, often on our tours. As a boy growing up in the midwest, he had engaged in numerous pranks, the most memorable to me being his mixing skunk juice or castor oil into the metal paint with which the school's radiators were to be coated. When the heat was turned on, the stench was, of course, unbearable, and resulted in his expulsion from school for a substantial time. I do not think his elders could have predicted that he would eventually become one of the most eminent educational statesmen of his generation.

Tyler traveled constantly, consulting to various agencies and governments on educational and social projects, with residences in different places for longer or shorter periods of time. Indeed, I thought of him as normally airborne, landing for meetings now and then. But he kept his wristwatch always on Chicago time, to avoid constant resetting at his different destinations. "I only have to ask where I am," he said, "and I know what time it is." He also retained his Chicago home as his main residence, and after he became a widower, employed a wonderful housekeeper, named Dolores, whom he entrusted with managing his home and many of the aspects of his work during his travels. Dolores was a religious woman and was much impressed with Tyler's forthcoming visit to the Holy Land for his meetings with our advisory committee. Tyler therefore arranged for her to join our group in Jerusalem as well, and she accompanied us on many of our tours.

Now Tyler had an unparalleled collection of anecdotes and jokes, many of which were off-color, and he regaled us with these in the intervals between serious talk. On one of our tours, Tyler began to tell one of his improper stories. As soon as he started, Dolores realized where he was headed and she said, "Dr. Tyler, if you're going to talk like that, I'm going to sing!" And she did, drowning out his story with her lusty rendition of a Biblical hymn.

The degree of his demand as an educational advisor to various agencies was brought home to us at the final dinner for our advisory committee on one of our Jerusalem visits. Tyler had had to leave a bit earlier than the rest of us, because he had to catch a plane to take him from Israel to Canada, where he was consulting on some educational project to McGill University. No sooner had he left our table than there was an urgent telephone call for him from the White House in Washington, but the call was too late to catch the flying Tyler.

One of the interesting things he told us was that one of his sons-in-law was an Israeli, the other an Egyptian. That must have made for many interesting family dinners, we thought. "How did you manage those occasions?" we asked. There was no problem at all, he told us, there was of course a good deal of discussion, but all of it was peaceable.

Tyler was invited by the Harvard Education School in the sixties to address the School on the new idea for a National Assessment of Educational Progress (NAEP), then being developed, of which he was a strong advocate. The audience of educators was strongly hostile to the idea, opposed to likely intrusions by bureaucrats into the educational process, fearful of invidious labeling of schools, apprehensive about increased reliance on standardized testing, worried about losses of funding, and so forth. I attended that meeting, curious to see how Tyler would fare in that den of lions eager to tear him apart. I needn't have doubted his capacity to handle himself. There he was, onstage, a short man with a mild manner, who seemed overwhelmed by the large and vigorous audience with the clearly adverse emotions. He began by calmly outlining the nature of the assessment, speaking in a slow clear voice, then waited for the storm of questions that burst forth. Each such question had obviously been anticipated by Tyler and worked through thoroughly in advance; each question, respectfully received, was answered coolly and fully. By the time the lengthy discussion period was over, Tyler had left no criticism unanswered, no question unaddressed. If he hadn't changed the mind of all his initial adversaries, he had nevertheless provided ample reasons for them to rethink their positions. And he had, in the process, defused the warlike atmosphere that had reigned in the room.

During the height of the student movement some years after, the Harvard School of Education invited him to come for a term as a visiting professor. The students were in a state of agitation and discontent, rebelling against traditional forms of university governance and inherited norms of scholarly research and discussion. Tyler was assigned some seminars to teach in the area of curriculum theory and development, an area of extreme ferment at the time. No doubt he would be seen by the students as an old fogy, a defender of the repressive university elite representing the dead hand of the past. As a matter of fact he sailed through his term, hearing the students out on every issue that was on their minds, and engaging them in a discussion of the issues that was genuinely educative, earning their respect and, indeed, their affection.

When, in 1972, I was invited to spend a year at the Stanford Center, of which Tyler was a major architect and first director, I came to appreciate in a new way his ingenuity as an educational administrator, even though he was no longer there, his term having concluded earlier. A large photograph of him graced the wall of one of the main rooms of the Center, and the tone he had set as the first director obviously still continued in evidence.

The Center invites a number of scholars to be in residence, typically for a year, chosen in such a way that there will be an overlapping of interests among major groups of fellows. Each scholar has a study on the Center's grounds, each is free to pursue his or her own research but since there is a central dining room, lunch is a time when the scholars naturally meet, across disciplinary lines, and when new ideas are broached. Voluntary associations are often formed through common interests and issue in weekly seminars or occasional conferences. The freedom, informal congeniality and beautiful California surroundings are highlights of the Center's features. But certain small touches seem to me inspired. One of these is the lack of telephones in the studies. The phones are located along an outside corridor, so that making a call requires some effort, and receiving a call as well. With the phone not at one's fingertips, a major distraction of ordinary academic life is done away with. I never had occasion to discuss this feature with Tyler, but I like to speculate that it was his genius that devised it.

On one of the last visits to Jerusalem of our advisory committee to the Hebrew University's Education School, we were standing in our hotel lobby after lunch before returning to our rooms. The lobby was pleasantly cool while the sun was fiercely bright and hot outdoors. As we began to go our separate ways, Tyler, dressed formally in black suit, shirt and tie, and hat, approached me to ask if I'd like to join him for a walk outdoors. It seems the doctor had advised him to take a walk every day for his health. I told him it did not strike me as such a great idea to walk outside at midday in the glare of the sun, especially dressed as he was. He smiled and said he was nevertheless going for his walk, and invited me once more to go along. "Thanks, Ralph, but no thanks", I said. And a fond memory I have of him is of his resolutely going out of the lobby cool into the sun's heat, calm, disciplined and with a twinkle in his eye as he left me. I saw him several times thereafter but that image captures, for me, some of the essence of Tyler's character and his charm.

Another member of Fox's advisory committee, already mentioned, was Joseph Schwab, of the University of Chicago Department of Education. Schwab, by training a biologist, was also an educator of shrewd judgment and wide learning in a variety of fields. Recognizing his talents, President Robert Hutchins of the University of Chicago had asked him to conduct a regular seminar of the College undergraduate faculty as a method of promoting intellectual integration at the College. Schwab had filled this role brilliantly.

In his teaching and his writings on education, he exemplified a broad Deweyan point of view, criticizing the tunnel vision of psychologists and social scientists who ride into the field of education armed with a single theory they consider the final truth. He himself championed what he called "the eclectic", the openness to different forms of inquiry to be employed intelligently in the effort to solve problems of practice. Correspondingly, he taught the virtue of respect for the various idioms characteristic of alien traditions of inquiry, and the need to learn and understand such idioms in order to profit from what they have to teach us. Further, he stressed the importance of translating the results of diverse inquiries into relevant educational contexts in order to design school curricula, rather than taking such inquiries neat. His writings on science education and his work on the "new biology" curriculum of the sixties as well as his approach to curriculum development were major contributions to educational thinking.

I had known of Schwab and read some of his essays (the titles of which had intrigued me: "Eros and Education",[1] "On the Corruption of Education by Psychology"[2]), before I met him. Our first meeting antedated by several years our joint membership of Fox's advisory committee. It took place, in fact, at a meeting of a small group of educators, of which we were both members, invited by the Presbyterian Church of the U.S.A. to review their plans for a new Sunday School curriculum, a review process which, as I recall, they regularly repeated every few years.

Having introduced ourselves, our group seated itself at one long table, and awaited our chairman's initial statement of the agenda for the meeting. Schwab was sitting directly to my right, a slight man with a determined look. We had all been given folders with documents to be considered in our deliberations, each such folder placed on the table before every seat. But Schwab's folder was flanked with a wide array of smoking materials he had retrieved from his jacket pockets, and neatly set out on the table as soon as he had sat down. There were a couple of cigars of regular size, some smaller and slenderer ones, a pack of cigarettes, two pipes, a tobacco pouch and a lighter.

As soon as the chairman began to speak, Schwab lit up, and he continued to smoke continuously during the whole of our two-hour meeting, starting with cigarettes, following with cigars of both sizes, concluding with alternating pipes continually refilled from his tobacco pouch. The general awareness of the dangers of smoking had not reached current levels until much later, but I have anyhow often wondered how a professional biologist, such as Schwab, could smoke so unrestrainedly. In any event, his contributions to our discussions at the meeting were so well thought out, so acute, and so helpful to our hosts that I left with a most favorable impression of his intelligence as a theorist of education.

It was not until many years had passed that our paths crossed again, when I received an invitation to visit the University of Chicago. It seemed the University wanted to explore the possibility that I might join their Department of Education. No doubt Schwab was behind this invitation, and he urged me to consider the possibility seriously. We had good conversations about the matter when I visited Chicago and he introduced me to some of his colleagues and showed me around the campus. There was much that was attractive about the University, in particular its hospitality to cross-departmental studies and its strong theoretical bent. In the end, however, I decided to remain at Harvard, where I already had put down firm roots. But the meeting with Schwab helped to confirm our friendly relations.

It was therefore a pleasant surprise to encounter him again when we both served for the first time on Fox's Visiting Committee in Jerusalem. Here, as members of the Committee, we had much more occasion for interaction and further opportunity to get to know one another. We met a number of the professors both individually and in seminars. We were asked to review many of the departments and several of the projects of the School of Education and to evaluate the work of a number of the faculty members, in particular, those who were candidates for promotion. Aside from these essential matters, we also spent much time in one another's company, on tours of the city and country arranged for us by Fox, and informally at mealtimes at our hotel.

Schwab, in these encounters, showed himself to be a complex and interesting person. Learned in a variety of fields, he was well versed in several scientific specialties outside of his training in biology, but he also had an informed background in history, literature, and religion—all tempered by an overriding concern for educational method. He considered himself an educator, not a philosopher, but he drew philosophical inspiration from Aristotle and Dewey in particular.

Pleasant in informal exchanges, he occasionally put people off by his curtness or impatience. On one occasion, our Committee had been convened for an evening meeting with various of the faculty members of the School of Education, among whom was a distinguished mathematics professor at the University. During the discussion, Schwab had occasion to make a comment, in the course of which he used the relatively obsolete word "diremption". The mathematics professor, whose native language was not English, asked Schwab what it meant. Instead of explaining the term or paraphrasing the statement in which he had used it, he simply repeated his statement with emphasis. When the professor ventured again to ask what it meant, he loudly exclaimed, "diremption , diremption!" At this point, the professor gave up, and slunk back in her seat, while the rest of us were simply embarrassed by his rudeness.

In 1971, Fox organized a conference on curriculum to take place in Jerusalem in conjunction with another meeting of our Visiting Committee. He invited me to give a main address, on the topic of Schwab's curriculum theory. Following the format at philosophy conferences, I wrote a serious paper appraising Schwab's writings on the practical as a focus for curriculum, starting with an exposition of his views and proceeding to offer several critical comments. I had asked Fox if he thought a critical review was appropriate and he assured me it was fine. Accordingly, I presented my paper in Schwab's presence, expecting that he would respond, either favorably or unfavorably, to my arguments. To my surprise, however, when called upon to reply, he said, "I cannot respond at this time", or words to that effect. I took his act to be an expression of honesty, indicating that he felt he needed to study my paper for awhile, unwilling to respond off the cuff before he had had the opportunity to do so. I hoped his act was not an indication that I had hurt his feelings by my critique.

His further demeanor reinforced my positive interpretation, for in ensuing contacts during those meetings and thereafter, he behaved in a perfectly friendly and relaxed way toward me. Indeed, after the meetings were concluded and we had all returned home, we struck up a friendly, if irregular, correspondence triggered by an interesting book that one of us had run across or some odd fact that seemed to us intriguing. Schwab was a great connoisseur of wine, as well as novels, and he sent me one memorable letter with an opinion of a certain wine I had asked him about (I forget which). He praised the wine but with the reservation that it had an aftertaste as if one-third of a mouse had decayed at the bottom of the bottle.

During the following year, when I spent my sabbatical at the Center for Advanced Study in Palo Alto, my wife Roz and I took a drive down the coast and visited with Schwab and his wife—also a Roz—in Santa Barbara where they lived. They were both altogether cordial. Schwab cooked dinner for us and the conversation ranged over a wide gamut of common interests, education, politics, literature, and philosophy. In the ensuing years, we continued our sporadic correspondence.

One of the projects I had completed at the Center was the preparation of a collection of my education essays entitled *Reason and Teaching.*[3] In that volume, I included my Jerusalem conference paper on Schwab, and sent him a copy. He wrote to thank me but, either in that letter or through the grapevine, I learned that he was miffed at my including the paper in my collection, where he had had no chance to reply, rather than submitting it to a journal which might have invited him to respond in the same issue. I was sorry that he was miffed but did not feel that I had in any way restricted his capacity to reply, in any venue that he chose. Indeed, I was gratified to receive a copy of his selected essays, *Science, Curriculum and Liberal Education* edited by his students Ian Westbury and Neil J. Wilkof, who took the occasion to include a critical response to my Jerusalem paper after all.

I have mentioned Lawrence Cremin in connection with the founding of The National Academy of Education in the sixties. But I had known Cremin earlier, having met him when he was invited to teach courses in the history of education at the Harvard Summer School some years earlier. It was Robert Ulich who had first mentioned him to me with strong approval as a young historian of education at Teachers College worth keeping an eye on. Thereafter, Ulich it was who promoted his visiting appointment at the Summer School. Cremin and his wife, and my wife and I hit it off immediately that summer. We were about the same age and our young children were also in the same age range. That summer, Roz and I in fact lent our playpen to Larry and Charlotte for the use of one of their children.

Since he and I were both teaching in the Summer School that year, we saw each other quite often and frequently had lunch together. An obvious bond between us was our common interest in American studies and in educational thought. He was at that time working on his history of progressive education, *The Transformation of the School*,[4] and I was lecturing on the pragmatists, both projects overlapping considerably and focussed in good part on Dewey and his influence.

Our special researches and methods of course diverged. He would, outside of class, disappear into the depths of Widener Library and bury himself for hours in newspapers of the era he was studying, emerging refreshed from this dusty pursuit and eager to get back to it as soon as possible. I was preparing my lectures and working on them as precursors of chapters in my book on pragmatism, which appeared in 1974. In addition, I was reading various publications on analytic ethics that summer, and I remember conversations with Larry about all these topics–the early progressive era, the pragmatists, and ethical theory as related to educational problems.

We learned a good deal from each other that summer and enjoyed also reminiscing about Columbia and talking about New York and Cambridge. There was a growing interest at the School of Education in the possibility of attracting Cremin to Harvard and although I do not know for sure, my best guess is that serious overtures were made to him in that vein. Cremin was however a true New Yorker, totally attached to the city and totally loyal as well to Teachers College, where he had studied and taught and where, in the event, he later became its President. He and I did however forge an informal but genuine alliance of interest and attitude that would issue in a growing connection between our two institutions, the Graduate School of Education at Harvard and Teachers College at Columbia, both dedicated to a broad view of education and the improvement of educational practice based on the highest quality of research and scholarship.

We both thought that educational studies had been too constricted, too focussed on local and contemporary school practice. We both wanted to retain such practice as a main focus but to place it within the largest framework possible, historical, philosophical, and comparative. In my teaching, I strove for such breadth through analytic and philosophical means. Larry studied the historical roots of contemporary problems and also emphasized the fact that education is not the sole property of schools but is channeled through social institutions of enormous variety, whose operations and effects deserve sustained study.

I have earlier noted that the initial idea of founding a learned society devoted to education came from Robert Ulich, who talked about it informally with Cremin and myself. Cremin eventually took the initiative and suggested to me that we should both try to put this idea to the test. A true researcher, he decided to gather the constitutions of several learned societies and see if we could not draft an appropriate parallel for an educational society to be thought of as a National Academy. On one wintry spring day, Larry came, with his batch of constitutions, to my home in Newton in the mid-morning and we set to work, fortified by a lunch of salami sandwiches on rye and black coffee that I had prepared for the occasion. Whatever else was accomplished that day, the sandwiches made a great impression on him, since on at least two occasions years later, when Larry gave public lectures at Harvard, he began by mentioning the legendary salami sandwiches which sustained us at the birth of the Academy, or better, its conception. For it took several meetings and numerous consultations before the National Academy of Education was actually born in 1965 and began the work it has continued ever since.

Meantime, relations between the faculties of Teachers College and Harvard continued to strengthen. For some years, we had joint meetings of the history and philosophy scholars of both schools, at least one at Teachers College that I can remember, and at least one at Harvard, where the topic I can recall was moral education, and where the participants included Philip Phenix of the College, John Rawls of Harvard, and R. S. Peters of London, then visiting at the Harvard School of Education. This regular form of contact could, however, not be sustained for long because of the complex schedules of the faculties involved, but the collaborative spirit remained strong through publication, personal communication and professional conferences. As the Academy itself took shape, its annual meetings fulfilled one of its most important functions, to bring Education scholars together regularly for discussions and exchange of ideas.

Further occasions for my meeting with Cremin took place when he joined Seymour Fox's external committee of advisors to the John Dewey Education School of the Hebrew University in Jerusalem. In the committee deliberations, there was ample opportunity for exchange of ideas on educational research and philosophy. In addition, Cremin and I met on at least one occasion in San Francisco in connection with my preparation for the year my wife and I would spend at the Center for Advanced Study.

During all this time, Cremin was making progress in his historical research, publishing papers and planning an ambitious project as well, a three-volume history of American education from colonial times until the contemporary period. Simultaneously, he was taking on heavy responsibilities at Teachers College in teaching and administration, first as head of the history and philosophy area of the College, then as head of the whole. Outside the College, he was active in the National Academy, and later as head of the Spencer Foundation for educational research.

When Ulich had first mentioned Cremin to me as a young historian worth keeping an eye on, he was prescient, but his judgment was an understatement. Cremin became a major figure in American education, not only as a historian of stature, but also as a national educational statesman. His energy was remarkable, combining a strenuous program of research, writing and publication with onerous duties of administration. How he managed these tasks remains a mystery to me, but his masterly three-volume history, *American Education* (1607–1980),[5] slowly took shape and began to be actualized. He told me how he had proudly presented his mother with Volume 1, which had take him years to complete, only to be asked, "And where are Volumes 2 and 3?"

His array of public activities and responsibilities were integrated into a framework which included precious time for family and friendship–for his wife and his children, and for a large company of friends with whom he kept in touch in various ways, large and small. My wife and I enjoyed his friendship over the years and on one occasion, amusing in hindsight, benefited from his friendly concern in his role as administrator. Teachers College had invited me to attend their Commencement, at which I was to be given an award for my work in philosophy of education. The invitation, which I accepted, was made several months before the scheduled June Commencement date, and Teachers College had informed me that they had made reservations for my wife and me at a small elegant hotel opposite the Metropolitan Museum for the occasion. We were to stay overnight at the hotel the day before Commencement, to take place the morning of the following day.

We arrived as planned at the hotel, where the desk informed us that we had no reservation. There must be some mistake, we said, since the reservations had been made months before by Teachers College. No, they said, there was no reservation for us and absolutely no room available at the hotel in any case. The manager, whom we asked to see, repeated that we had no room reserved and that, in any case, none was available. The most he could do was to refer us to another hotel and to pay our taxi fare to get there.

Having no alternative, we drove to the other hotel and were ushered to our room. This room turned out to be appalling. It looked as if it hadn't been used for months; it was dusty, dirty and altogether unsuitable. At this point, not knowing what else to do, we left our bags in the room and found a public telephone in a shop nearby where we phoned Larry Cremin. He couldn't believe what had happened. Then, thinking for a moment, he suggested we visit the Guggenheim Museum for a few hours and phone him again after we had seen their exhibits. I told him I was sorry to trouble him with this problem, but he said not to worry. "Just phone me after you visit the Guggenheim."

Roz and I did as he had suggested and after an enjoyable tour of the museum, phoned him again. This time all he said was, "Why don't you just get your bags, return to the original hotel and see the manager." But, we said, it was the manager himself who had told us in no uncertain terms that the hotel had no available room at all. "Just go back anyway and ask for him," said Larry.

We returned, still not believing anything could be done, and asked for the manager. He emerged from his office, all smiles and full of apologies. There had been some mistake, he assured us unctuously, and led us to a room. The room turned out to be not a room but a suite, with living room, bedroom, dining room and kitchen, all beautifully furnished and decorated, spotlessly clean, and with a bowl of fresh fruit on the table. So much for managers' assurances. We had a wonderful stay and the next morning told Larry what had happened, thanking him and again apologizing for putting him to the trouble. "Don't apologize", he said, "I enjoyed myself enormously. I've always had a craving to tell a hotel manager what I thought of him."

Nobody had any concern about Larry's health. He was thin, athletic, never smoked, ate and drank in moderation, etc. When, several years later, he died of a sudden heart attack, just having arrived at Teachers College as usual one morning, his death produced a wave of shock and disbelief in those of us who knew him. It is wonderful that he was able to commit so much of his work to writing, but his death was a serious blow both to history and to education.

CHAPTER 10

THE SCHOLAR AND THE ANALYST

I got to know Harry Wolfson shortly after my arrival at Harvard, having been introduced to him after lunch one day at the Faculty Club. He had his lunch every day at the Club, where he presided over what came to be known as Wolfson's table, where he was often joined by several Harvard scholars including Jakob Rosenberg of the Fine Arts Department, the legal scholar Samuel Thorne, my predecessor at the Education School Robert Ulich, the Divinity School professor George H. Williams, the philosopher Morton White, and occasional others. He welcomed me warmly when we met, expressing his interest in the field of education and his pleasure in the fact that I had had a strong background in Jewish studies, and had graduated from the Jewish Theological Seminary of America. He was, thereafter, cordial whenever we met, urging me to promote the advancement of Jewish education in Boston by helping develop a school or schools to that purpose. It was all very vague, but exceedingly friendly. Years later, his typical remark to me when we would see one another in the lobby of the Faculty Club was "We must talk sometime", which was never followed by an actual meeting.

Harry Wolfson was a rare and formidable scholar, a loner, whose whole life was dedicated to philosophical, theological, and historical studies of the Greek, Jewish, Christian, and Moslem heritages of thought. Born in a small town in Lithuania in 1887, he came to the U.S. in 1903 and was admitted to Harvard in 1908 on a High School Scholarship of $250. There he remained all his life, first as instructor in Hebrew literature and philosophy, eventually as Nathan Littauer Professor of Hebrew literature and philosophy, the first chair of its kind in the United States or Europe, having been promoted for that position by the then foremost Christian historian of Judaism, the great Harvard scholar George Foot Moore. As a young man, Wolfson conceived the major theme of his work–the influence of the Jewish philosopher of the first century, Philo of Alexandria, on the whole subsequent course of Western philosophy. His grand plan was to develop this theme by writing a series of books ranging over the history of philosophy from the pre-Socratic philosophers through Spinoza. He began by writing books on Crescas' critique of Aristotle, on Spinoza, and on Philo. Later came books on the Church Fathers, on the Mohammedan Kalam, on various problems in religious philosophy, and various manuscripts remaining after his death in 1974, at the age of 87.

So voluminous and authoritative was his scholarship that he was frequently, it was said, mistaken for a committee. Indeed, on one occasion when I had lunch at Wolfson's table along with Edmund King, an English visitor whom I had invited to join me, I introduced him to Wolfson, briefly listing some of the books he had authored, whereupon my guest turned to Wolfson and, with obvious appreciation for the work, asked "And who are your contributors?"

The range of Wolfson's work necessitated his mastery of many languages. One apocryphal story concerns a graduate student who came to consult him about a possible topic for a thesis proposal. Wolfson heartily approved the topic, clambered up the ladder in his study to retrieve a relevant reference book, which he held out to the student, saying "You read Phoenician, of course?"

His mastery extended also, and obviously, to the most esoteric branches of Talmudic literature. I recall visiting him once, on a rare occasion, in his Widener Library study, while he was with an eminent guest, the Jewish legal scholar Professor Daube, originally from England and at that time from California. They were discoursing, in my presence, on various sections of the Talmudic literature, and hit on the topic of the ancient Temple sacrifices, a much neglected area of study. They both waxed enthusiastic, eager, as it seemed to me, to revisit the minutiae of rules and customs dormant for two millennia. Pulling various books off the shelves, they were soon in a world of their own, gleeful as children playing with a new toy.

Wolfson's knowledge of the many languages required by his research did not treat them as self-enclosed regions. The philosophical ideas he treated spread, after all, across numerous language barriers and required tracking across such barriers. Wolfson therefore concerned himself with parallel expressions and other meaning relations connecting the different languages functioning as philosophical vehicles.

He was meticulous in mapping such relations and also in deriving the philosophical force of every nuance in expressing parallel concepts applying across languages. Indeed, the one sort of criticism that I have heard about his interpretations of philosophical works is that he leaves no room for ordinary happenstance in the process of writing, for example, ordinary variation in word choice, slips of the pen, stylistic or aesthetic predilections, even ignorance or forgetfulness about what other writers have said, occasioning all too human lapses from doctrinal consistency. Instead, every work emerges from his hands with every nuance doctrinally significant, every such work totally consequential in its own terms. The picture seems, indeed, too good to be true. Yet, in his defense, it might be said that it is just too easy to ignore such nuances as insignificant for interpretation unless, as is typically not the case, there is independent reason to suppose that the deviation in question resulted from a particular lapse evident in the circumstances. In any case, the philosophical interpretations Wolfson produced had to survive an unusually stringent set of tests, since their consequences ramified widely across many different cultures, philosophical traditions, and languages and constituted so many different hostages to fortune.

It is worth special notice that the scholarly arena within which Wolfson worked at Harvard is so catholic, and so unified in its pursuit of truth that a scholar's own cultural identity is of little consequence in deciding the realm of his interests or the probity of his results. Thus, as I have mentioned earlier, the Christian scholar

George Foot Moore was a renowned authority on the history of Judaism. After he brought the Jewish scholar, Wolfson, to Harvard, Wolfson turned out to be one of the main authorities on the philosophy of early Christianity. Indeed, when his book, *The Philosophy of the Church Fathers: Faith, Trinity, Incarnation,*[1] appeared, Harvard held a special convocation at the Divinity School to celebrate the event. Several years later, when Wolfson's book on the Kalam, the philosophy of medieval Moslem scholasticism, appeared,[2] Harvard hosted another meeting to mark the occasion. This was an event to which Moslem scholars from across the world were invited, including those from Arab countries which were radical foes of Israel. A large group assembled for this celebratory occasion, including several eminent Arab scholars who stipulated that their participation was contingent on its not being reported in the press.

Wolfson was of course especially valued at Harvard as a unique scholar of unusual breadth and depth. One occasion I remember vividly was the Alfred North Whitehead lecture Wolfson gave at Harvard on the topic, "Descendants of Platonic Ideas", in which he traced the twists and turns of the problem of universals from its beginnings through its medieval forms to modern times. The main moral of the lecture was that the historical richness and complexity of the problem were virtually unknown to contemporary philosophers, who commented on it but were ignorant of its roots.

The chairman of the occasion was Professor Donald Williams of the Harvard philosophy department, who introduced Wolfson to a packed audience in the large lecture room of Emerson Hall. Williams began by citing several of Wolfson's publications and confirming for the record that he was indeed a single scholar and not a committee. Wolfson then proceeded to give a brilliant account of the vicissitudes of the problem throughout its history, concluding with his criticism of contemporary philosophers unaware of this history. The questions then came quick and fast, to all of which Wolfson responded, in his East European Jewish accent, with obvious gusto, pleased at the audience's interest. One question toward the end of the discussion period has remained in my memory over the years. The questioner asked Wolfson something that pertained not to the problem of universals but to a different issue relevant to the medieval period. With great good humor, Wolfson replied, approximately as follows, "Why do you ask me about something I did not say; why don't you ask me about something I did say. It so happens that, on the topic you raise, I have written 7 papers since 1934", and then he proceeded to answer the question.

Wolfson's personal life is the subject of many stories, and no doubt many exaggerations. He lived in an apartment not far from the Widener Library and it was said that he was the first one entering the library in the mornings and the last one to leave, a saying I believe not far from the truth. He was certainly a loner, not in the sense that he could not be sociable. Indeed, he conversed easily with people and his manner was indeed charming. It is rather that the world of books and scholarship was so much more attractive to him than the world of social interaction that he spent the greatest part of his life in the former world and was jealous of any time taken away from it.

He enjoyed his daily social lunch at "Wolfson's table" in the Faculty Club, but that was only a small part of his day, preceded and followed by many many hours of scholarly research and writing. He talked easily with strangers to whom he had just been introduced, typically querying them about the origins of their family names or the historical background of their immediate ancestors, but he would not prolong such conversations, eager as he was to get back to his books. He hardly ever accepted dinner invitations, or invitations to social occasions of whatever sort and would typically decline offers of transportation to his destination by friends whom he encountered while on some errand. My wife and I once met him leaving a movie one evening and offered to give him a ride in our car to his home, but he refused, preferring to take the subway instead.

His moviegoing was the source of many Wolfson stories. He did, apparently, often attend movies at the end of the day. It was said that cowboy movies were his favorites, although the picture he had seen when my wife and I met him and offered him a ride was a remake of the original Blue Angel, with Marlene Dietrich. His method of watching a movie was to be attentive for the first ten or fifteen minutes, then to lapse into a deep sleep until he awoke shortly before the end. His theory was that if you know how a movie begins as well as how it turns out, it is relatively easy to reconstruct the middle.

His scholarly monasticism did not eliminate his need to hear the human voice, at lunch, or at the movies, or even by accident. The following incident was told me by a friend of mine who had broken through Wolfson's reserve to the point of visiting him at times at his apartment. During one such visit, according to my friend, the phone rang. Wolfson picked up the receiver, listened a while and, after saying, "Sorry, you have a wrong number.", hung up. He then explained to my friend that his phone number, through some similarity to the number of a local shop, had occasioned his receiving many calls intended for that shop. My friend then suggested to Wolfson that he telephone the phone company and request that his number be changed. "No," said Wolfson, "I am glad to get such calls, and to hear a human voice."

That his all-consuming dedication to the labors of scholarship gave him a rather skewed perspective on worldly affairs and human motivations is illustrated by a story relayed to me by a former student of his many years after the fact. Reportedly, Wolfson said to him, not long after the Russian revolution, something like this, "Lenin, Trotsky, Radek and the rest are all intelligent men, but they are wasting their time. If they would only sit down with a text, they might accomplish something."

His total immersion in scholarship seemed to me also to produce a rare conscientiousness in his relations to colleagues. This trait was brought home to me as the result of a visit to Harvard from New Zealand by the young Spinoza scholar Edwin M. Curley. Having lunch with him one day, I asked if he had met Wolfson yet on his visit. He was totally surprised, having assumed that Wolfson, whose book on Spinoza had been known to him for years, was no longer among the living. I asked him if he would like to meet Wolfson, an offer he eagerly accepted. The next time I saw Wolfson at the Faculty Club, I told him a young Spinoza scholar visiting from New Zealand was eager to meet him. Wolfson invited me to bring him around to his Widener study one day.

A few days later, I took Curley with me and knocked on Wolfson's door in Widener. He welcomed us cordially and, after the introductions were over, the conversation turned to Spinoza. Wolfson asked about Curley's work and then the two of them entered into a discussion of Wolfson's early book on Spinoza and the later scholarly literature. Wolfson took out a large file drawer with index cards containing annotated summaries of virtually every paper and book on Spinoza published since his own book had appeared. At this point, not being a Spinoza scholar myself, and finding their discussion too abstruse and remote from my own interests, I let my mind wander while they engaged in intensive discussion of the works referred to by several of the file cards. They continued in this vein for another half hour or so, when I felt my presence was no longer needed and, besides, I was finding it more and more difficult to avoid dozing off. So I excused myself and left, while they continued their avid discussion of Spinoza studies. I felt a glow of self-satisfaction at having done a good deed in bringing these two scholars together, and thought nothing more about the encounter.

Some weeks later, I ran into Wolfson at the Faculty Club and said, "Hello, Harry". To my astonishment, he responded by refusing my handshake and saying, "I'm very angry at you", then turning away. When I followed him and asked, "Why are you angry?", he said, "I can't talk about it", and walked away. I was baffled but there was nothing I could do but wait for my next encounter. When I met him again some weeks later, the same scenario was repeated. I said hello, he said he was very angry at me but could not talk about it. This happened two or three times more in the succeeding months.

About a half year later, when we met in the Faculty Club and he again repeated that he was angry at me, I did not let him escape. I took his arm and drew him over to a sofa in the Club lounge. I said, "Harry, you've got to tell me why you're so angry at me." He was reluctant but I was insistent, and he finally was able to express his feelings in the matter. "Because of you", he said, "I allowed myself to criticize other scholars in front of that young man you brought to my study. I would never have done so otherwise, but because you were in the room and I know you, I let myself make critical remarks about the published work of other students of Spinoza in the presence of a young scholar".

"That's it?" I said, relieved. But he was dead serious, feeling that he had shown insufficient respect for scholarly colleagues by his informal criticisms, thus setting a bad example to a younger man, and damaging his self-image as a careful and judicious scholar. He had of course published very stringent critiques of colleagues over the years, but I assume he separated such formal published criticisms from off-the-cuff comments in conversation, which may have appeared to him to verge on irresponsible gossip. This finely honed conscientiousness struck me as exceedingly odd and certainly rare. But in an age when loose talk abounds in academic exchanges, I believe his countervailing conduct makes a valuable point and exemplifies an ideal that ought to be more widely adopted.

The diversity of scholarly interests and methodological presumptions is one of the glories of research universities. Sometimes such diversity generates new ideas and unanticipated strategies of inquiry. Sometimes it engenders only an appreciation of alien worlds of scholarship with no practical effect on one's own. At times it does not even do that, representing diverging scholarly initiatives that move past each other silently like vessels that pass in the night. In the latter case, only the outside observer may view the diverging vessels with admiration and understanding, valuing their separate destinations and itineraries. Such was a chance encounter between the scholarly world represented by Wolfson and that represented by J. L. Austin, which I was able to observe.

The late J. L. Austin was the most eminent of the Oxford philosophers of the 1940's and 1950's, one of the architects of the "ordinary language" philosophy that flourished in that period and beyond in English-speaking countries. His version of this philosophy was distinctive in that it based itself on close study of the Oxford English Dictionary and in this sense was founded on fact rather than speculation. He did not, like Wittgenstein, construe philosophy as a solely therapeutic enterprise aiming to relieve the mental cramps caused by particular abuses of ordinary language. Rather, he thought of philosophy as a systematic empirical study of the fine structure of English meanings, capable of serving as an antidote to the premature generalizations, based on a few choice examples, that bedevil traditional philosophizing.

His papers, "How to Talk: Some Simple Ways", "Other Minds", and "A Plea for Excuses"[3] dazzled his contemporaries when they first appeared, and his later books, *How to do Things With Words*,[4] *Sense and Sensibilia*,[5] and *Philosophical Papers*,[6] published after his untimely death in 1960, consolidated his stature as a brilliantly innovative philosophical thinker. In Oxford, he was quickly recognized as a leading philosopher and feared, as well, as a formidable critic.

When he first visited Harvard, I attended his seminar on the topic of excuses and could see for myself how incisive and original he was. He explained his topic by saying that philosophers had paid almost exclusive attention to nouns and verbs, whereas he was going to deal essentially with adverbs, notably those such as "inadvertently", "by accident", etc. which served to diminish the full responsibility for an action, thus constituting excuses. He then described certain actions and invited the participants to describe them adverbially or to choose between different adverbial descriptions as most appropriate. There ensued seminar discussions in which participants proffered their descriptions and discussed them or proposed related cases where similar choices were elicited. At the end of each session, the general distinctions and emphases he had wanted to bring out were in fact brought out, no matter how the discussion wandered. His seminars, I thought at the time, were like the operation of fate. It didn't matter what road was taken, you ended always at the predestined location.

The William James lectures, *How to do Things with Words*, which he gave at Harvard, set out, in systematic fashion, his main topic of "performative utterances", which constitute actions proper rather than their descriptions. Thus, "I promise to return the book" doesn't describe a certain promise; it is in fact the promise in question. This topic which had concerned him as a quite young philosopher in 1939, was considerably elaborated in these lectures as a "theory of illocutionary forces" in 1955, and had an enormous influence as the origin of so-called "speech act" studies. In 1960, he died, at the age of 49, a true loss to philosophy.

Austin presented an intimidating appearance to American students. His dress was certainly more formal than these students were used to, in the British style, dark suits, white shirt and a tie, and a black Homburg. Together with his philosophical reputation, he seemed unapproachable. Yet his conduct belied his appearance. He was available to student initiatives, of one or another sort. He accepted invitations to address student fora, lecturing at least once, to my knowledge, to the classics club on Aristotle, and attending end of year student parties where ball games were played, (in which he participated), amidst the general conviviality.

One of my Masters students at the School of Education took his "Excuses" seminar for credit, which was brave of her; most of those in his crowded seminars were auditors afraid to risk a low grade, only about 6 or 7 enrolling for credit. A year or so later, my student decided to apply for admission to the Education doctoral program and needed to supply letters of reference. I asked her if she had requested such a letter from Austin who had given her an *A* grade. She was shocked at my suggestion, thinking it altogether out of the question, since the seminar had taken place a long time ago and he would not remember her; in addition, she was a lowly Education student, not a philosophy student and only a Masters level student as well. How could she dare to write him back at Oxford, requesting a letter? I encouraged her to throw her doubts to the wind and send an air mail request to Austin. Within ten days she had a reply from him; he not only remembered her but also recalled the quality of her work and wrote a glowing recommendation in support of her application.

He had developed a warm acquaintanceship with my late colleague, Professor Roderick Firth of the Harvard philosophy department; they had been engaging in a department-sponsored series of debates on issues of mutual interest, about which they had quite opposing views. In the Spring, Firth invited Austin to his home for dinner in Lexington. It was a fine day when Austin arrived, in his usual formal attire. The Firths took him into their garden for drinks before dinner. After chatting for a while over drinks, they excused themselves and returned to the house to complete their dinner preparations, having first asked him if he would mind waiting in the garden. He had said that he would certainly not mind and would in fact enjoy the garden on that lovely day. Their conscience eased, they busied themselves with preparing the dinner in the dining room of their house, leaving Austin sitting in the garden chair outdoors, taking pleasure in the surroundings. After about a half hour they came back into the garden but Austin was not in his chair and was not visible at all in the main garden area directly in back of their house. Puzzled, they walked around their grounds and discovered him in an area to the side. He had removed his coat, hat, and tie and, having spotted their mower earlier, was now energetically engaged in mowing their lawn.

During that first visit of his to Harvard, when I attended his seminar, I sent him a couple of reprints of my early papers, written in quite a different vein from his characteristic work. He responded by inviting me for a walk. We strolled around Cambridge for a good while. He appeared to be genuinely interested in my work and we chatted for some time about it. His manner was open, informal and friendly and this first walk led to others later on, in which he revealed some of his thoughts about philosophy, both his own and American varieties, thoughts which revealed a practical, mischievous, and quite unusual side to his imagination. About Norman Malcolm, an American devotee of Wittgenstein and his host in the U.S. I remember Austin's comment "Who but a Nebraskan farm boy could have introduced that Austrian princeling to Americans."

As to his own philosophy, he indicated his strong belief that his work needed to have an empirical basis in information on the actual speech habits of English speakers. How can such information, relating to speakers scattered widely both geographically and sociologically, be gathered? Austin suggested that there are large numbers of retired people with nothing interesting to occupy them, who could be employed in administering questionnaires to speakers all across the land and on every socio-economic level. A marriage made in heaven. On the one hand, scattered bits of information difficult to retrieve; on the other hand, an army of bored elderly people equipped with questionnaires to gather such bits, knowing that they were helping to advance science in so doing and reviving their morale in the process. A ludicrous scheme, I thought, but interesting in revealing Austin's empirical bent, which was rather unbelievable to Oxford colleagues, for whom conceptual analysis and empirical studies were sharply separated, philosophy being concerned solely with the conceptual realm. In this respect, Austin broke rank with his colleagues, defending the relevance of empirical studies to philosophy, a relevance much more congenial to American than to Oxford philosophers at the time.

As to American philosophy, what exercised him as an urgent practical problem was the need to bring philosophers together. In Oxford, there were over fifty philosophy dons in the same university, all discussing philosophy in continuous contact. In America, as he found on his lecture trips particularly, philosophers were isolated. A couple of idealist philosophers in one university, three or four physicalists in another city, nine or so realists in still another and so forth. The ideological differences between them could indeed serve to enrich philosophy if only they could be in closer contact, as in Oxford. Austin's suggestion was to arrange telephone conferences at frequent intervals when all the American philosophers could argue some issue in intellectual if not physical proximity. This idea was indeed not ludicrous and in fact was ahead of its time, when teleconferencing has become a common occurrence. Whether it can ever approximate the live debates and discussions of Oxford does seem to me, however, dubious.

When the time came, at the conclusion of Austin's second visit, for him to return to Oxford, the philosophy department arranged a farewell party for him. Wolfson, as member of the philosophy faculty, was invited as a matter of course, but no one expected him to attend, since he never attended the department's functions. Moreover, Austin's work was so far removed from Wolfson's that there seemed no obvious philosophical bond between them. Austin was wholly concentrated on articulating the concepts and distinctions exemplified by the vernacular of the ordinary English speaker, untainted by theory or doctrinal predilections, whereas Wolfson's whole life's work was devoted to the analysis and interpretation of doctrines, philosophical and theological.

To everyone's surprise, Wolfson did come to the party, ate and drank and chatted pleasantly with whoever came within his orbit. Toward the end of the evening, guests took seats around the central tables; a few people then made farewell remarks in honor of Austin, and wished him a bon voyage. When these remarks were over, the guests, in the fine afterglow of the occasion, did not immediately depart but lingered and conversed with their nearby neighbors. As it happened, Austin was seated next to Wolfson and I was seated in a chair or two over.

Austin was engaged in animated philosophical discussion with some of his nearby neighbors; Wolfson, however, had nodded, sleeping peacefully while holding a glass of scotch that he had been sipping. He was oblivious to the thrust and parry of the philosophical argumentation raging around him until, at one point, the word "God" was mentioned by one of the interlocutors. Wolfson immediately came awake, sat upright in his chair and, turning to Austin, said, "Do I understand you are interested in theology?" To which Austin, with obvious disdain, replied, "No, I am not interested in theology. I am interested in religion." Wolfson then sighed, "Oh, you are not interested in theology?", and immediately went back to sleep. Two illustrious thinkers, inhabiting different worlds, had passed each other noiselessly in the night, leaving it up to us, their hearers, to appreciate them both.

Wolfson was a loner, but he needed to hear the human voice. He also needed approbation even though he was supremely self-confident about his own scholarly views. These characteristics converged in an odd and touching way the last time I visited him in his study, at his invitation.

I found him in a cheerful frame of mind and the reason soon became clear. A large handsome volume lay on his desk. It was the product of many hands, beautifully illustrated with photographs and art work, and containing essays by various scholars, including a lengthy contribution by Wolfson himself. The book, recently published, had just arrived at his study a day or so before and he had been engrossed in poring over it.

He was eager to show it to me, to have me share in his pleasure. I riffled through the pages, obviously impressed. The focus of the book was the history of Christianity and I seem to recall that Wolfson's contribution dealt with themes relating to the doctrines of the early Church. I expressed my admiration of the book, thanked him for inviting me, and prepared to leave. But he restrained me, and asked me to be seated and stay a while.

When I sat down near his desk, he put the book on my lap, turned to his own essay, and asked me to read it. I started to do so but he interrupted my concentration. "No," he said, "I want you to read it aloud." I did not comprehend his intent, but I complied, starting at the beginning and reading his essay aloud to him. He beamed with satisfaction, stopping me after a couple of paragraphs, interjecting with questions to elicit my admiring comments. "Do you see?" "Isn't that right?" and questions of similar sort kept punctuating my reading every few sentences.

After a couple of pages, I indicated to him that I had to leave. He was obviously disappointed and seemed ready for me to continue indefinitely in the same vein. The situation seemed to me sad. His solitary bookish existence did not fully satisfy, nor did his well-earned conviction of the lasting value of his scholarship, nor even the public acknowledgements of his stature in publications and special convocations. In his solitude, he still craved a voice to respond personally with approval, even if it were a voice articulating his own words, reflecting his self-approbation back to him. That was the last encounter I had with Wolfson, not the renowned scholar but the human being.

CHAPTER 11

CENTER FOR ADVANCED STUDY

I took my third sabbatical, in 1972–3, at the Center for Advanced Study in the Behavioral Sciences, located in Stanford, California. The Center, as I knew it, was an unusual institution, which invited about fifty fellows each year to pursue their researches at its campus adjacent to Stanford University. The designation "Behavioral Sciences" was interpreted quite broadly, to include not only psychology and the social sciences but also such areas as linguistics, history, psychiatry, computer science, medicine, literary theory, and philosophy, among others. The effort was made to invite for each year a cohort of scholars whose disciplines, though spanning different fields, might reasonably be expected to relate to one another in fruitful ways.

The physical appointments at the Center were superb. On a rise overlooking the university grounds, the Center buildings were afforded a beautiful vista of the Stanford campus, its tiled structures separated by grassy expanses, trees, and connecting paths. The Center itself had a central cluster of buildings, one housing administrators and support staff, another for secretarial and typing assistants, and a third with a large meeting room and adjacent lunchroom. At the approach to the central cluster was a large loquat tree—a surprise since, as an Easterner, I had never seen or known of such a tree with its succulent fruit.

From the central cluster, walkways led gently down a slope toward the individual studies assigned to the fellows, each study offering a stunning view of the university grounds. A wide space adjacent to the central cluster contained a number of rustic tables and chairs where the fellows could take their lunch if they chose not to eat indoors in the lunchroom. The whole arrangement was designed to facilitate interactions among the fellows in a pleasant, informal atmosphere. Additional amenities included a messenger service, which was available to take book requests from the fellows every day and to try to retrieve the requested books from the Stanford library, or to borrow them from other libraries if not available at Stanford. All in all, an idyllic environment.

No one was required to meet any deadlines, attend meetings, or file any reports. The initial invitation to be considered for a Center fellowship required a response including a plan of research, but beyond that preliminary response, no one stood over the fellows' shoulders to monitor how the plan was being carried out or if, indeed, it had survived the interval from response to eventual arrival at the Center. Such freedom was daunting, to say the least. To be given such ideal conditions for work with no external institutional demands constituted, for the great majority of the fellows, an abrupt change from the busy atmosphere of the universities in which they normally worked, with their heavy rounds of duties, deadlines, and demands. The Center's freedom, however, imposed its own pressure and exerted its own paradoxical threat. What was one going to be able to show for one's efforts at the end of the year, given all these goodies?

Most academics require some time to get used to a new study, even in familiar surroundings, before being able to set to work. I. A. Richards once told me he needed to stare out the window for several weeks after moving his desk to a new room before being able to concentrate. And, as seemed to be the case at the Center, those fellows who arrived with no momentum due to a well-worked out plan had a good deal of difficulty in getting started, with every passing hour of idleness adding to the pressure felt. Those with an advance plan, especially one already realized in part, could set to work after minimal physical orientation and with considerably less uneasiness.

The physical orientation, especially for a city-bred Easterner like myself, was radical, startling as well as exhilarating. The climate of California itself was a constant wonder, sharply at odds with my past experience in New York, Philadelphia, and Boston. The glorious sunny weather of the Center I experienced as a precious marvel. "How do you manage to do any research or writing in such fantastic weather?" I asked a Californian colleague. "You have the wrong conception," he replied. "As an Easterner, you think of good weather as a rarity that has to be enjoyed while it lasts, because it won't be here for long. But Californians like myself know that it will be here waiting for us no matter how much time we spend in our studies or in dusty libraries. We can afford to be profligate with it."

My surprising encounters with the California setting involved more than the splendid weather. On my first work day at the Center, I parked my car bright and early at the Center's lot and started to walk down the path leading past the central administrative building on my way towards the study that had been assigned me. As I was passing the glass door of the administrative building, I stopped short in some dismay. For there, about two and a half yards ahead of me on the path, I saw what seemed to me a huge spider moving toward me methodically at a slow, even pace. I sprang to the side, opened the door and, trying to appear calm, called to the nearest secretary, "Deanna, would you see what's on the path outside the door?" Deanna opened the door, took a look outside, and said nonchalantly, "Oh, that's a tarantula." She then returned to her desk, picked up a Styrofoam cup and an index card, and went outside as I followed, eager to see how she would deal with this natural emergency. She placed the cup horizontally on the path with the opening directly facing the monster. Veering neither to the right nor left, the oblivious tarantula continued its march right into the cup, whereupon Deanna picked it up, covered it with the index card and, returning inside, placed the cup on her desk. "What are you going to do with it?" I asked, baffled. "Oh, I'll take it home for my son; he has a terrarium and will be happy to have it." It was only later, after some research in the Encyclopedia Britannica, that I learned that tarantulas, unlike scorpions for example, are not really dangerous and that their bites are not usually fatal.

After the tarantula encounter, a second surprise awaited me that morning. Having proceeded finally to my assigned study, I was charmed by its location and furnishings, including a large desk, filing cabinets, bookshelves, and a glass door that gave onto a small patio overlooking the Stanford campus below. I explored these pleasant surroundings and knew that I would enjoy working here. Opening the main drawer of my desk, I discovered two sets of room keys, one of them obviously an extra, which I decided to return to the administration building. Back at the building, I found Deanna still at her desk. "Guess what I found in my desk drawer," I exclaimed. "A snake?" she asked, hardly looking up from her typing. "No, an extra pair of room keys," I replied weakly, mightily impressed by the possibility of finding snakes in my desk. I never in fact encountered any snakes in my room during the period of my fellowship, but after a while did get used to seeing a particular snake, presumably harmless, patrolling the paths and grounds of the Center at various times of the day. I came to think of him as our friendly Center mascot.

The year at the Center turned out most pleasant and productive for me, not least because of the various scholars I met and with whom I was able to exchange ideas regularly, both formally and informally. The special benefit of such exchanges was that they cut across the disciplinary boundaries that normally enclosed us in our respective university settings. An additional benefit was that they overrode the typical preconceptions of location—in my own case confronting me not only with the California style but with the diverse backgrounds of fellows who hailed from other parts of the country as well as from other countries.

One of the fellows with whom my wife and I soon became friends was the late Robert Heizer, who was at the time, I believe, curator of the anthropology museum at Berkeley, named for A. L. Kroeber, with whom he had studied. Soon after the fellows had all arrived at the Center, we were assembled for a plenary meeting where we could introduce ourselves to our new colleagues and say a word about our own spheres of interest. Following this meeting, we all retired to our studies to begin our work.

About a half-hour later, I heard a knock on my door. Opening it, I saw a slender, medium-tall man with a short beard, clad in a sweater and jeans, holding a large book. This was Bob Heizer. I had seen him at the plenary meeting but had not exchanged any words with him. After mutual introductions, he held out the book for me to glance at. It was a handsome volume, with full-page color photographs of rock paintings of the Chumash Indians of the Southwest. Heizer went on to explain the purpose of his visit. "You're a philosopher," he said, "and therefore don't know any facts but can presumably discern meanings. I want you to examine these paintings and see if you can figure out what they mean. I'll leave the book with you and we can talk about it later." With that, he departed.

I was to learn that the Chumash were an illiterate people who left brilliant paintings on rocks in various locations in California and elsewhere and that Heizer had traveled throughout these areas to find and photograph these paintings. The book reproduced them and gave detailed information about their placement and measurements. Several of the pictures, because of their symmetry, seemed to me to depict insects or other living forms; others exemplified regular geometrical shapes. But the majority differed from both these sorts. They were similar, in my eyes, to abstract expressionist works of the recent past. What meaning they might have had for the Chumash I had no idea and had to report this result to Heizer at our next meeting, when I was revealed as someone not only bereft of facts but also apparently oblivious to meanings.

Our further conversations nevertheless went well and indeed flowered into a warm friendship. Roz and I much enjoyed hearing about his interests and his work. He was a tremendously diligent scholar and prolific writer on a wide variety of subjects pertaining to culture. A formal lecture he delivered at the Center, based on a research paper of his in *Science*, was entitled "Heavy Engineering in the Ancient World" and was an offshoot of his intense interest in megaliths around the world, for example, in Sicily, Easter Island, and Egypt. The lecture explained how the huge rocks could have been moved from their points of origin to their eventual faraway places in the ancients' constructions, and how they might have been maneuvered into their desired positions; it also inferred the sort of social structure necessary for such ancient constructions to have been accomplished.

Heizer told us about his son, Michael, of whom he was justifiably proud. Michael had achieved recognition as an environmental artist, whose work consisted in altering the landscape and photographing the result. On one occasion Heizer told us about he hired a tractor with a suitable attachment and used it to create a trench in the Nevada desert, forming a straight line and then turning it to the left to form an obtuse angle, finally photographing the result from the air. On another occasion, he took an enormous boulder aloft in a helicopter, then let it drop onto a stretch of the desert while he snapped the sandy splash. As Bob showed my wife and me some of the photographs of Michael's work, it was hard for us to resist the thought of a classical Freudian duet being played by father and son, the one studying order and construction, the other spontaneity and destruction, Dad setting them up and junior knocking them down.

Both displayed an evident initiative and independence of mind. Bob, as we got to know him, embodied what seemed to us a surprising combination of traits—the steely autonomy of an adventurer and the meticulousness of a bookish scholar, the tough guy attitude he displayed to the world at large and the empathetic understanding he showed for downtrodden native peoples of the Americas.

The tough guy exterior was no mere pose. He told us that on one of his trips to Sicily to see the megaliths, he had been followed by a thug who seemed interested in the camera equipment he was carrying. He tried unsuccessfully to shake him until he decided to slow sufficiently to let the thug get close. "And then?", I asked. "I turned and laid him flat," was the reply.

His dedication to scholarship was unremitting, as illustrated by the great number of Bob's published works—books and articles ranging from theoretical and methodological treatises to monographs on particular themes to polemical essays and focussed scientific studies. On our occasional visits to his local residence near the Center, my wife and I would invariably find him engrossed in writing a pencilled draft of some new article or book chapter.

Much of the scholarly work that occupied him had to do with the native peoples and their plight in America. He was full of rage at those who had oppressed them, missionaries who had without understanding mocked and uprooted their religions, languages, and cultural traditions, political leaders who had justified taking their lands and confining them to reservations, military men who had subjugated them by might, and American society at large, which remained content with their enforced isolation behind a wall of indifference, doomed to poverty and miserable conditions of life. In discussing these matters, his toughness, scholarly understanding, feelings of empathy and indignation were all melded into one.

Bob was not knowledgeable about philosophy nor was he interested in learning about what was going on in the field. Roz's background in clinical psychology and psychoanalytic modes of thought was closer to his domain, but did not really draw him out. Thus, we had essentially nothing professional to teach him, but we learned a great deal about the anthropological topics he was involved in, partly from our many discussions about his work and partly from a number of his books and papers which he gave us. And he evidently enjoyed our company as we enjoyed his, sharing a broad humanistic outlook. Thus we had, early in the academic year, begun to meet every so often for dinner, sometimes at a local restaurant, sometimes at our home, sometimes at his. These occasions were invariably pleasant and stimulating, the conversations animated and wide-ranging, the mood upbeat.

Toward the latter part of the academic year, we began to sense a subtle change in Bob, so subtle we could not be sure it was not a figment of our imagination. He seemed to us somewhat preoccupied, perhaps a bit less engaged, but we could not put our finger on it. A month or two after we first felt this delicate alteration in him, he phoned us at home one afternoon. "Can I come over this evening?" he asked. "Sure, Bob, look forward to seeing you!"

He arrived at about 8:00 o'clock, carrying a large carton. "I wanted you to have this," he said, opening it to reveal a case of Medoc wine, his favorite. "What's the occasion?" we asked. "I just wanted to be with you this evening, and I know we've enjoyed this wine together and wanted you to have it." Toward the end of the evening, he apologized to us for inviting himself over, intimating that he had seen his doctor that day and gotten some bad news. The change in his mood these past weeks was now explicable, the case of wine carrying the ominous suggestion that he would not live to enjoy it and so wanted to bequeath it to his friends while he still had the chance.

After that evening, Bob seemed to recover his former energy somewhat, working constantly in his Center study and engaging in discussions with colleagues. We resumed our occasional dinner meetings with him, which went on as before, though with an undertone of concern that was never brought to the surface. He never said another word to us about his doctor's report.

During the last week before the end of the Center year, we arranged a final dinner with him at our home. The conversation went very well, with no further reference to his health, but with unavoidable acknowledgment that this was indeed our last time together. As he turned to leave, he said, "I have something for you," and he gave us two items with obviously personal meaning for him. One was a large color photograph he had taken himself in Egypt, of boats on the Nile. The other, for my wife, was a circular piece of amber that he had collected somewhere, that might have been a pendant or amulet or part of a necklace. We thanked him for the gifts, grateful for these physical tokens of his friendship and, as well, for what he had taught us, not only about anthropology, but also about how one ought to live in the face of mortal danger. Several months after our departure from the Center, we found Bob's obituary in the Times.

One of the delightful features of the Center was that there were no mandatory meetings. Informal associations formed naturally and within a few weeks after the beginning of the academic year, voluntary seminars took shape spontaneously, based on interest. One such seminar took in the historians among the year's cohort and decided to focus on the theme of slavery. The group to which I belonged was named "The Fuzzy Language Seminar" and included a number of colleagues, all concerned in one way or another with informal aspects of communication and inquiry, but approaching this theme from quite different directions.

One of us, Steven Marcus, was a professor of literature, an expert on Dickens and on Freud, at the time completing a book on criticism called *Representations*.[1] Another, Joseph Weizenbaum, a professor of computer science at M.I.T., was working on a book to be called *Computer Power and Human Reason*.[2] A third member was Herbert Weiner, a psychiatrist who had been doing neurological research in the area of brain science and was completing a large work to be called *Psychobiology and Human Disease*.[3] I was myself working on my book *Four Pragmatists*,[4] at the time concentrating on Peirce's semiotic. Occasional visitors included a biologist, John Platt, and a German psychoanalyst, Alexander Mitscherlich, formerly head of the Sigmund Freud Institute in Germany. Mitscherlich had been in Switzerland during the Second World War, was later an observer of the Nuremberg Trials and co-author of a book on Nazi medical experiments, entitled *Doctors of Infamy: The Story of the Nazi Medical Crimes*.[5]

Our seminar met once a week during our Center residence and our discussions drew on and ranged over a good many of our interests—literary and scientific language, psychological and neurological approaches to the mind, functionalism and behaviorism, computation and reasoning, brain and thought, as well as several other topics. Marcus brought his literary background to our discussions, and he, along with Weiner, contributed a Freudian perspective as well. Weiner, at the time editor of the journal *Psychosomatic Medicine*, brought a medical perspective to bear, as did the occasional participant Dr. Alexander Mitscherlich. My own slant derived from my background in philosophy of science and philosophy of language as well as my work on Peirce and ongoing analyses of vagueness, ambiguity and metaphor which later issued in my book, *Beyond the Letter*.[6]

Weizenbaum brought expertise in computer science to bear, but far from being a single-minded technician and advocate, he was at a turning point in his thinking. He had earlier been one of the acknowledged pioneers in developing computer systems and, when I had first met him socially at one of Henry Aiken's parties many years before, he had displayed a supreme confidence that the computer could theoretically do anything a human being could do. "Could it recognize an occasional user?" I asked at that time. "Certainly," he had replied, "the keyboard could be fitted with pressure-sensitive keys, so it could record and thereafter recognize the individual pressure pattern of any user's fingers. The second time any given user, say Jim Jones, typed anything in, the computer could be set to respond, 'Hello, Jim, how have you been since we were last in touch seven days ago?'" The computer could, in Weizenbaum's view at the time, replicate any human process of thought or problem solving in any sphere.

At the Center, this view of his had already changed radically. Shortly before coming to the Center, he had published an article expressing his fear of the consequences of using computer systems in social institutions. His argument was straightforward: every such system has occasional breakdowns; it requires periodic inspection and, if necessary, "debugging." The bigger the system, the more likely are its breakdowns and the more time is required to detect its bugs after its breakdowns. Further, the more a system is assigned the making of decisions that are time-sensitive, the less likely that it can be debugged before the deadline for decision. Thus if military decisions, for example, are given over to the computer for execution in times of urgency, disaster may result—and the same holds for other classes of decision.

Weizenbaum had been worried, not only about the increasing size of the computer systems being built into social institutions, but also about the quality of the decisions such systems were increasingly being asked to make, and the seductive appeal they had for the public as presumably scientific, hence foolproof. He had, early in his career, worked on introducing computer systems into banking but, beginning to dread the extension of such systems into more personal areas of life, he had invented a counseling system, to serve as a horrible example.

This latter system was named "Eliza," and it was programmed to respond to any user seeking psychological counseling, as might be illustrated by the following hypothetical dialogue: Eliza: How can I help you? User: I'm at a loss. Eliza: What is your problem? User: I can't decide whether to marry my boy friend or not. Eliza: Tell me more. User: I'm of two minds. Eliza: Why do you say that? User: Some of my friends think I should not marry him but I'm in love. Eliza: How long have you felt this way? User: For at least a year. Eliza: Have you talked to your mother about this? Etc., etc.

The seductiveness of this ingenious program was evident to many people who tried it out at M.I.T. as soon as it became available. Even those who knew how it worked came under its spell and found it hard not to think of it as another person. Indeed, the program gained a certain notoriety as a computer "counselor" that could actually help people needing psychological help or advice. This sort of publicity was dismaying to Weizenbaum who had invented it precisely to show how dangerous it was since, although he thought it obviously useless in counseling, it was able to deceive people into thinking they were getting genuine advice from an informed interlocutor.

Weizenbaum's book, completed during his Center residence, provided a general exposition of principles of computing, emphasizing its power. But his book went on to disavow the claim of computers to provide a solid basis for human reasoning and decision-making. The two halves of the book's title, *Computer Power and Human Reason* were meant to be taken not as allied but as contrastive. His main argument was that in personal matters, in legal or political deliberation, in ethical and social issues, computers ought not to be given the role of decision-makers. Summing up his view in conversation, Weizenbaum put it this way: computers can be given deciding roles in all matters except those involving judgment or love.

Considered something of a rebel or a turncoat by his erstwhile allies in the computer field, he was roundly criticized for a failure of nerve by the rosy optimists of computing who saw utopias dancing before their eyes. But Weizenbaum, undeterred, gave as good as he got, serving as a much-needed gadfly to the field. In our seminar, he taught us all a good deal about computing while he gained support from the several humanistic fields we represented—literature, psychiatry, and philosophy.

The psychiatrists brought a special aura to our group. Personally astute, they were perennially shrewd and skeptical about people, alive to personal undertones and overtones and sensitive to nuances and hidden meanings. These characteristics jibed with Marcus' literary antennae as well as my own preoccupation with vagueness, ambiguity and metaphor. They also brought a certain relaxed outlook to their roles at the Center, which seemed to me to contrast with the nervous tension that characterized many of the fellows.

The pervasive pressure that I alluded to earlier, stemming from the freedom of our situation, was evidenced in the work habits of most of the fellows, or at least in the regular appearance of work. With the exception of lunch times, study doors were largely shut and the clattering of typewriters could be heard all through the grounds. Certainly, fellows conversed or strolled in pairs irregularly at various hours of the day and fellows could occasionally absent themselves from their studies to take trips or attend conferences elsewhere. But the regular tapping of typewriters was a constant accompaniment and a standing reproach as well as an incentive to those who at any hour knew in their hearts they were not steadily advancing in their projects. The situation reminded me of the jingling of coins continually being expelled by winning slot machines in a casino, spurring everyone to keep playing, while instilling envy and hopelessness in the losers. Those who came to the Center with a project already started could escape most of the pressure and I was glad to be one of those. But there were a number for whom the pressure was severe enough to profit from the informal friendly help offered by one or another of the psychiatric fellows in the class.

In their own demeanor too, the psychiatric fellows tended to deviate from the norm. One of the psychiatrists in the class previous to mine was Bruno Bettelheim, Director of the Orthogenic School at the University of Chicago. I had visited the Center one day during his residence and had lunch with him. When lunch was over and the diners all began to return to their studies for their writing, Bettelheim said to me, "Everyone thinks you're supposed to write during your fellowship year. I think you can very well read, and I have done nothing but read all year."

Dr. Mitscherlich had a different plan. When all the fellows left the lunchroom and returned to their studies for work, he went back to his study for a siesta. With complete aplomb, he followed this routine daily. Undisturbed by the noise of typewriters testifying to fellows hard at work in the surrounding studies, Mitscherlich would for two hours each afternoon, sleep the sleep of the just. An inspiring example to all of us, none of whom had the courage to emulate him.

In his calm way, he offered to help me in a peculiar predicament I found myself in a couple of months into our residence. Would that I had only taken his advice! The circumstances were these: I had been invited by the sociologists at Berkeley to give a lecture at one of their formal colloquia. Having accepted their invitation, I was informed of the arrangements they had prepared. The lecture was to take place in the evening, prior to which the Sociology Department was to host a small dinner in my honor at a nearby restaurant.

The Mitscherlichs and the Weiners had indicated their interest in attending the occasion and Alexander had offered to drive us from Palo Alto to Berkeley. Accordingly, on the appointed day, he and his wife, Herbert and Dora Weiner, and my wife and I all got into Alexander's BMW for the hour's drive to Berkeley. The trip began well and we were all in good spirits. However, at about a quarter hour into the drive, I began to experience a vague discomfort that developed into a full-fledged case of nausea. My wife became concerned, having noticed that my brow was perspiring and my face had turned white.

Her concern soon became evident to the rest of our company, whereupon both physicians, Drs. Mitscherlich and Weiner, urged me to cancel my lecture and turn back. But I refused, finding it impossible to break my promise at this late date. Alexander then proposed to stop at the nearest pharmacy and get me some Gelusil-lac, which he said would surely cure my discomfort. I continued stubbornly to refuse, while he insisted that I allow him to stop. "I'm a doctor," he said "and I can prescribe medicines for you."

My stubbornness prevailed, however, and we drove on without interruption, arriving at the restaurant at the appointed time. The rich aromas of spices within announced it as a Greek restaurant, the savory cuisine whetting everyone's appetite except mine, for I wished only to get into fresh air without eating at all. The chairman as well as others of his department who had joined us for dinner all greeted me most cordially and engaged me in friendly conversation as we seated ourselves at table. When the tangy appetizers arrived, my wife now recalls my having turned green. My task was to engage with a smile in the continuing repartee, just nibbling at the crackers and trying hard not to faint.

Finally, the ordeal over, we left the restaurant and drove to the auditorium for my lecture. To add to my consternation, the auditorium was packed, and I was conducted to the stage where a podium stood ready. Before turning to the podium, however, I took the moderator aside and asked her to point me toward the exit and the bathroom, should I need to leave hurriedly in the middle of my lecture. Having done so, she strode to the podium and introduced me.

I then replaced her at the podium and for the first time looked at the audience in dread and took my paper out of its manila envelope. It was a very warm evening and I was feeling flushed anyhow. So, before beginning my lecture, I asked the audience to excuse me for removing my jacket, which I did, thereafter removing my tie as well, unbuttoning my collar, and rolling up my shirtsleeves. In a warm cloud of nausea, I began my lecture, prepared at any moment to cut and run if I needed to. My voice was low and I had to concentrate hard on keeping it loud enough to reach the farthest seats, thankful that a microphone had been prepared for me.

My typescript was about 30 double-spaced pages long and at about the tenth page, my nausea seemed to moderate. My surprised relief continued to grow until the end, to the degree that when the question period arrived, I felt almost completely normal. With the give and take of the ensuing discussion, I rose to the top of my form and ended the occasion elated that I had survived without disgracing myself. The residual emotion I had was sincere regret that I hadn't taken Mitscherlich's advice en route and so saved myself the cliff-hanger experience that followed.

There was a sequel to this story in which my other psychiatrist colleague, Herbert Weiner, played the leading role. Soon after my Berkeley lecture, I consulted a physician at Palo Alto about the likely cause of my gastric distress. After what seemed to me a long and rigorous physical examination, he assured me that there was nothing wrong with me, in particular no gall bladder problem, and that I ought simply to take antacid pills to control my symptoms. Reassured, I bought a supply of Gelusil tablets that I continued to chew regularly. After our year at the Center was up, my wife and I returned to Boston to attend our son's graduation, then left for a summer's trip to Europe with our daughter, before coming home for good.

Visiting the great museums of London, Paris, Amsterdam, and Brussels, we were enthralled by the magnificent works of art in each of these cities. But I was never without an ample supply of Gelusil tablets, and my memories of the museums are to this day suffused with recollections of intermittent nausea and heartburn kept at bay by the tablets I was continually chewing in the elegant corridors. I resolved that when we returned home, I would fix a consultation with my own physician to seek definitive relief from my continual distress.

He did in fact examine me not long after we got home, and repeated the diagnosis of my Palo Alto doctor, assuring me I had no gall bladder problem and approving my antacid regimen. During a phone conversation about a week later with Dr. Herbert

Weiner to catch up after our departure from the Center and our return to regular routines, he asked me how I was. Instead of taking his question as a *pro forma* ritual of politeness, I proceeded to describe my symptoms, whereupon he responded by asking if I had been tested for gall bladder trouble. I said two doctors had assured me that I did not have such trouble. He replied quietly, "But have you been tested?" I said, "No." He then concluded gently, "Why don't you ask your doctor to give you the test?" I followed his advice and phoned the next day to make an appointment. After the test, I was told the results would be available in a few days.

I was seated in my office about a week later when my phone rang. It was my own physician. He said, "I've just gotten your test results. Boy, I wouldn't want to have what you have! You have several huge gallstones and they are the cause of your trouble." Since this was the same doctor who had, without suggesting I take the test, assured me that I certainly had no gall bladder problem, I concluded it was he who had the more serious problem of plain gall.

I eventually did have my gall bladder removed—as a result of telephone advice long distance from a psychiatrist! Herbert Weiner and I since remained fast friends. When his book *Psychobiology and Human Disease* appeared, I realized from its pages how seriously he took the interaction of body and mind. When philosophers even today hear references to the mind-body problem, they reflexively think of Descartes. I learned from Weiner to think rather of stress, gout, ulcer, and sudden death, a contribution to my philosophical maturity.

CHAPTER 12

SOME INVITED LECTURES

One of the benefits of academic scholarship is the opportunity to interact with colleagues outside of one's own university. Such opportunity is always available through the medium of books, professional journals, and correspondence, and regularly available at conferences and meetings in one's disciplinary area. But invited lectures offer a richer occasion for intellectual and human interchange, since they typically require a longer period of contact with colleagues and the chance to savor the ethos of another university department, which is greater than the sum of its members. I have already (in Chapter 8) described my pleasant time in Oberlin when I delivered the Mead-Swing lectures there in 1965.

After my return home from the Center at Stanford, I accepted an invitation to give a series of three or four lectures over a period of a week at Notre Dame University in Indiana. I had never been to Notre Dame, but had met the philosopher of science Father Ernan McMullin of the Notre Dame faculty at philosophical conferences and enjoyed our occasional conversations. I was also interested in seeing how a religiously sponsored university might reconcile such sponsorship with its commitment to free inquiry.

The terms of my invitation were thus quite congenial to me. In addition to delivering my several lectures, I was encouraged to acquaint myself with the university at large, and also to visit, as well as participate in, seminars offered by the philosophy department. My hosts were exceedingly cordial, and I enjoyed seeing McMullin again and meeting the other members of the department. One of my lectures dealt with American pragmatism and I was interested in discussing it with Cornelius Delaney, the resident scholar of this philosophical movement. Other lectures touched on philosophy of science, McMullin's specialty, and formed a focus for some of our individual conversations.

The Catholic culture of the university was quite evident throughout. The campus itself declared its religious affiliation, its two central paths perpendicular to each other laid out in the form of a cross. Christian symbols decorated various parts of the university structures, and crucifixes adorned the walls of the formal guest rooms. The side of one of the main buildings displayed a huge mosaic image of Jesus, with one hand raised in benediction, and a student informed me that the image was deliberately placed so as to overlook the football field, enabling the Notre Dame team to receive a Divine blessing before home games.

The team loyalty is a well-known Notre Dame tradition, and I encountered one aspect of its intensity when I checked into my room at the Faculty Club. In talking to the desk clerk, I found out that the dates for my lectures had been arranged so as not to coincide with any scheduled home game. The reason, I was told, is that every Club room is owned, in perpetuity, by a Notre Dame fan—for every date on which a home game is played. I was also told that the scholarly Father Theodore Hesburgh had intended, when he acceded to the presidency of Notre Dame, to reduce the football fervor of the university in order to bolster its academic image but had been quickly disabused by his trustees.

In any event, the academic quality of the seminars I took part in was excellent. The epistemology seminar, devoted to the work of my late Harvard colleague, Roderick Firth, was indistinguishable, as far as I could tell, from comparable discussions anywhere in its familiarity with the relevant analytic literature and the wide-ranging development of its themes. Analogously, the philosophy of science class conducted by Father McMullin was, I believe, as universal and open in its treatment as any comparable class in any secular university. My conclusion is that, whatever aspects of religious sponsorship are ingredient in the structure and functioning of Notre Dame, they do not constrain the academic freedom or quality of its philosophy offerings. I was impressed by the way general standards of academic scholarship were able to live peacefully with the particular religious culture of the university. I take it as an object lesson of the harmony achievable between cultural traditions and objectivity.

When I arrived for my week's residence at Notre Dame, I was suffering from back spasms and had difficulty walking without pain. However, the winter weather set in a day later, and brought me a pleasant surprise. When it dipped below zero, my spasms moderated. That evening, my philosophy colleagues invited me to dinner at a building across campus. The night air was icy but my long walk to the event gave me total relief. The conviviality of the occasion added to my good mood, the humorous sallies and witticisms flowing easily all evening. One particular anecdote is still fresh in my mind, told by a senior professor of the department now deceased. It seems he often visited Italy with his wife and brought back works of art, which are free of customs duty on arrival in this country. On one such trip, he brought back two packages, one a small painting by an Italian painter whose name I don't recall but let's say it was Reni. The other package was an Italian cheese; let's suppose it was a Bel Paese. As they neared the customs area in the U.S., his wife reminded him that the cheese needed to be declared. He reassured her, saying, "Leave it to me." He handed the customs official the package with the painting first. "What's that?" asked the official. "It's a Reni." "What's that?" "It's a work of art." "Okay, now what's in the second package?" "It's a Bel Paese." "Also a work of art?" "Certainly." "Fine, you can go."

I was scheduled to give my final lecture in the early afternoon of my departure day. The lecture room had a large glass window at the front of the room, looking out on a lovely stretch of the campus grounds. The audience was seated in several rows all facing the window, while I stood at the podium with my back to the window, facing the audience.

The lecture began well, the audience appearing friendly and receptive. As it continued, I noticed some frowns here and there and some signs of nervousness which I couldn't interpret. There was nothing to do, however, but to keep on. So I continued, surer than ever that the listeners were uncomfortable or displeased or upset. I began to worry that I must have offended them somehow. There were numerous Catholic clergy in the audience and so, while I kept on reading from my typescript, my mind at a second level of consciousness started to review what I had read so far to detect any remarks that might have struck them as religiously offensive. My search proved fruitless but the experience was disconcerting and I was glad finally to reach the end of my lecture. Unaccountably, this finale drew enthusiastic applause, completely at odds with the unhappy atmosphere that had prevailed earlier.

I was totally puzzled until I left the podium and looked behind me through the window. Apparently, soon after I had begun my lecture, it had begun to snow. As I continued, the snow had kept falling at a rapid pace, picking up ominously as I neared the end. The displeasure of the audience was due, not to what I was reading, but to the growing snowstorm that they could see but I could not. I realized, too, that they were concerned that my flight, which I had planned to take as soon as I could get to the airport, might be delayed or cancelled.

I was myself apprehensive and did not know whether I could actually get to the airport. However, my hosts assured me they could drive me there, despite the snow, and told me that on analogous occasions their guests had managed to make their flights. Several of the clergy further added that they would pray for me. The story in fact ended happily; I managed to get on the last outgoing flight before the airport was shut down, returning home with warm memories of Notre Dame.

In 1981 I received an invitation to give a paper at an October conference of the German Semiotic Society, to take place in Hamburg. I assume that I was invited because of my involvement as advisor to the Peirce Edition Project in Indiana, as well as my books, *Four Pragmatists*[1] and *Beyond The Letter*[2]. At any rate, I was glad to accept, thinking I might encourage a connection between semiotics and analytic philosophy of language, which had occupied me in the latter book primarily. When, however, I offered the Society the option of my giving a talk on recent developments in American analytic philosophy or my giving a general talk on Peirce and pragmatism, they chose the latter.

The meetings themselves reinforced my growing impression that the members of the Society, as well as the whole field of semiotics, were mesmerized by Peirce and were more involved in interpreting and expounding his doctrines than in pursuing new investigative leads. The most I could do at the conference thus was to offer a paper critical of Peirce rather than another laudatory account. As one who has been active in the Peirce Edition Project to produce a chronological edition of the master's work, as well as the Peirce Society and the John Dewey Foundation to boot, I appreciate, and indeed applaud, organized efforts by succeeding generations to conserve, rethink and extend the ideas of important philosophers. But I have often felt such organized efforts to be vulnerable to uncritical idealization of their heroes, and insufficient attention to new inquiries inspired by alien sources.

Be that as it may, the Hamburg conference provided impressive evidence of the wide knowledge and deep understanding that German scholars had of Peirce. Unlike the American philosophical scene, the interest in Peirce evident in Hamburg seemed to permeate many more academic branches and intellectual specialties than was the case in the United States, where American thought was pursued by no means as widely as English and continental varieties.

At Harvard, for example, the focus of philosophical attention had been centered for decades on Kant, while Peirce and Dewey received scant attention, primarily in a course or two taught by Morton White and myself. By contrast, a German scholar attending a Harvard meeting I arranged some years after the Hamburg conference startled me by remarking how disappointed she was to have met a number of our graduate students all working on Kant, when they might better be occupied in devoting their studies to Peirce.

No doubt, the devastation of German intellectual life in the Nazi period accounted for some of these differences. American thought had been re-encountered or discovered afresh after the Second World War and kindled new hopes for democratic life especially in the younger generation. Many of the graduate students I met in Hamburg were already quite sophisticated Peirce scholars who, unfortunately, because of the German university system, could not hope for academic appointments; they instead produced translations and scholarly studies while they supported themselves by non-academic pursuits.

The wide interest in Peirce by professionals in Hamburg also struck me as novel. Moreover, it was not only sociologists and film makers, naturally concerned with symbols and images, who showed such interest, but also physicians and medical researchers, whose similar interest was a surprise to me. For them, the link to Peirce was through the concept of "symptom", a semiotic function in clinical diagnoses and classifications. Indeed, the professionals' applications of Peirce seemed to me the most intriguing in stretching the boundaries of his semiotics even while taking his fundamental categories for granted.

In 1989, an international Peirce conference, on the occasion of his 150th birthday, was held at Harvard. Here too, Peirce scholarship predominated while a number of us attempted to relate Peirce to new directions of thought, inspired by the belief that Peirce himself would have been interested in such directions, had he been present, rather than in revisiting his own ideas of over a century ago. But here again, the balance I have mentioned earlier, between conservation and exploration, was tremendously difficult to achieve. Conserving the funded achievements of the past is surely important in itself but such importance becomes evident only in light of its meaning, which continues to show itself in its sequelae.

Conferences, such as the one at Hamburg and the one at Harvard, do much more than exchange ideas. They equip ideas with persons who defend them and oppose them, exposing the live emotional, social and political potential they possess, only minimally discernible in the bare words in which they are couched. By bringing the individual scholar into contact with a wide array of persons outside his local milieu, they broaden his native perception of an idea's potential and so deepen his understanding. Seeing what the Hamburg attendees saw in Peirce, I think I gained a new appreciation of some potentials of his ideas.

The city of Hamburg itself was an emotional experience for me. Once installed in my hotel room by my cordial host, Professor Klaus Oehler, I left the hotel to stroll around its environs, overwhelmed by the knowledge that Hitler's hordes had marched down these very streets, looked into these same elegant storefronts, broken into some of these very houses to murder their Jewish inhabitants or send them off to the crematoria. I knew that those days were gone, that the government was democratic, that the large majority of the youth were liberal and trying to come to terms with the Nazi evil of the past. But history is not so easily dismissed. The unspeakable savagery of just decades earlier hung over my view of the city and would not be expunged.

My mood lightened at the conference sessions, especially by encounters with the youth, who struck me as serious and scholarly in demeanor and liberal in political orientation. The faculty I met were exceptionally pleasant and open, but I found myself, almost without thinking, calculating their ages to speculate on what they were doing during the war.

A welcome surprise was the bust of Ernst Cassirer, in the entryway to an important university building where he had presumably taught before he had had to flee. I had heard this gentle and learned scholar lecture on Rousseau and Kant in New York at the New School during the 40's and had, much later, read his *Language and Myth*,[3] *Essay on Man*[4] and *Philosophy of Symbolic Forms*.[5] To imagine this humane intellect as having to flee for his life from the brutal nightmare of Nazism was almost unbearable, the emotion somewhat softened by the vision of his bust in the entryway, evidence of his ultimate victory over the barbarians in the recovery of sanity by his native country.

Several of us at the conference were moved to pose for a group photo in front of Cassirer's bust, among whom, as I remember, were Tom and Irene Winner, Klaus Oehler, Tom Sebeok, and myself. Klaus Oehler had translated *Pragmatism* by William James, and later sent me his translation of Aristotle's *Categories*; we kept in touch for a while through the Peirce Society. The Winners and I remained in close contact in the States, first when they invited me to give a talk at their Semiotic Institute at Brown and thereafter through their affiliation with my Philosophy of Education Research Center at Harvard.

Tom Sebeok was a major promoter of semiotic studies at Indiana and I saw him on occasion at meetings of the Peirce project there. An active academic administrator, he also became well known for his book denying animal language. I had some interesting though inconclusive conversations with him, but I remember his lively sense of humor, in particular his definition of four medical specialties: A surgeon knows nothing and does everything; an internist knows everything and does nothing; a psychiatrist knows nothing and does nothing; a pathologist knows everything and does everything, but it's unfortunately too late.

One afternoon in 1987 I received a phone call at home. The voice was unfamiliar but was quite articulate and clear; it sounded, indeed, as if it had originated just a few blocks from my house. But the caller introduced herself as Francesca Gobbo, a professor of education in the University of Padua. She was in fact phoning from Italy to invite me to deliver four lectures on education at her university.

Startled and elated at the prospect, I quickly expressed provisional acceptance, pending further study of the details. Soon thereafter, I phoned her to convey my agreement to come, and my wife and I set about making our travel plans for later that year.

Our flight was uneventful but our landing was unplanned. Padua itself has no suitable airport; the nearest one being in Venice, where we were due to arrive at the appointed time. Because of an airport workers' strike, however, our plane was diverted to Treviso shortly before landing. So, without being able to alert our hosts, who awaited us in Venice, we arrived instead at Treviso, Padua being situated between the two cities, with which it formed a triangle.

Having gathered our suitcases and emerged from customs, my wife and I had to decide how to proceed. We decided to try to get to Padua where we knew the name of the hotel at which accommodations had been reserved for us. Still unable to communicate with our hosts, who would soon no doubt realize what had happened, we confronted a new problem, unfamiliar as we were with the airport and the transportation system, and almost totally ignorant of Italian to boot.

We located some directories and maps and determined there was a bus to Padua, but where to pick it up and how to purchase tickets? Having first purchased some Italian currency at the airport bank, we managed to locate a clerk whose rudimentary English matched our rudimentary Italian, supplemented with copious hand gestures and inquisitive facial expressions. She indicated that bus tickets to Padua could be bought at the bar across the street and that our bus stopped in front of the bar.

We thanked her and proceeded to half-carry and half-wheel our suitcases across the busy street while dodging the traffic, ending up finally at our destination, the bar. While my wife waited outside with our luggage, I entered upon what I suppose was a typical scene: a goodly number of local men having drinks at the bar, served by a busy bartender intent on his work. I had finally to catch his attention by hand waving, whereupon he approached, looked me in the eye and waited for me to speak up. From some unknown depth of my unconscious, the words came, unbidden: "Due bigletti a Padua, per favore?". Expressionless, the bartender reached under the bar, retrieved two tickets and handed them over. I held out my hand in return, containing all the Italian change and bills I had. He picked out what he needed and I was on my way, flushed by my linguistic coup.

We were lucky, in boarding the bus, which arrived soon thereafter, that two strapping teenage passengers near the door sprang to our aid in getting our suitcases up into the bus and in getting them down about forty minutes later at our destination. We were also fortunate in locating our hotel not far from our stop and arriving in the lobby without further incident, though unavoidably later than our originally scheduled arrival time. There we were surprised and delighted to find Professor Gobbo who, having understood what had happened, had driven hurriedly from the Venice airport in time to welcome my wife and me on our arrival at the hotel. Thus began a warm friendship between us that continues to this day.

We found Padua to be a charming university town which was a joy to explore on foot, including lavish open-air fruit and flower markets and a spacious circular park. But the high point for us was the small Scrovegni chapel, the interior of which was covered by spectacular frescoes of Giotto. Entering the chapel, one was immediately surrounded on all sides by biblical scenes of dominantly brilliant blue, a magic world of its own.

The university itself, founded in 1222 as an offshoot of the university of Bologna, housed the lecture hall where Galileo had taught, and was now a fully modern institution hosting, during our stay, an international medical conference on the uses of platinum and other metals in the treatment of disease. My lectures were to begin a couple of days after our arrival and Professor Gobbo suggested meeting the very next day to consider the translations she had prepared of my texts, which I had furnished earlier.

The translations proved to be very instructive. I had been prepared for the procedure of lecturing with the aid of translations by my participation in a project several years earlier, organized by the late George Bereday of Teachers College Columbia. The project invited several Japanese teachers to spend several months in the United States, living with American families and working in local public schools. The first week after their arrival, these teachers were given a week's orientation at the Experiment for International Living in Vermont. My role was to provide two or three lectures on American philosophy during the orientation period.

My instructions were to lecture, using words of no longer than two syllables, and sentences containing no more than ten words; I was also to pause after every two or three sentences, to allow the translators time to translate efficiently. Despite these constraints, all went well for a while, the four Japanese graduate students serving as translators proceeding in regular fashion without apparent difficulty. Suddenly, however, the whole process stalled, the translators apparently baffled as to how to go on, carrying on an animated discussion amongst themselves in Japanese until they managed to agree as to how to translate my last comments. The trouble, as I learned later, turned out to be my use of the philosophically commonplace English word "mind", which opened itself to too many possible equivalents in German and French, alluded to by my translators, and also in Japanese.

The translators later told me this story: William Faulkner was giving a lecture in Japan, translated in process to the audience. Asked later by his translator how he had felt about the whole process, he said he was quite pleased but mystified by one point. While he was telling an extended joke, the translator had allowed him to proceed to the end without pausing at all for the translation, which was no longer than the usual short string of words he had become accustomed to. Yet the audience had gotten the point, laughing loudly at the punch line. "Very simple to explain", said the translator. "I let you tell your anecdote without interruption. Then I said to the audience, "Mr. Faulkner tells joke. Laugh!"

Still, my discussions with Francesca Gobbo about her translations carried various surprises. The main one having to do with my theme was an inversion of the two terms "education" and "pedagogy" in Italian educators' usage as compared with the American. For Americans, the term "pedagogy" is the more practical of the two in its connotations, having to do with procedures of teaching and schooling, whereas the term "education" is the more theoretical, connoting systematic doctrines and principles guiding pedagogical practice. For Italians, according to Gobbo, the situation is the reverse, "pedagogy" being the more theoretical in its connotation, "education" being the more practical or procedural. The result of this inversion meant that she had had to go through a fair amount of recasting of my text in order to get the intended meaning across to the Italian audience.

The presentation of my lectures proceeded as in the Japanese case mentioned earlier. I read a short passage in English, Gobbo translated it into Italian and thus we continued until the whole was finished. In the discussion period, questions were raised in Italian, translated for my benefit into English; I replied in English which was translated back into Italian. The lectures thus took quite a bit longer than the usual lecture time, but were, I thought, well received. Many in the audience knew some English but were too diffident to phrase their questions in English; the process thus worked out best for all.

My lectures were given in one of the old lecture rooms of the university, which impressed upon me the long tradition of the university itself. In my opening comments, I elaborated on this point. In my own country, I said, Harvard is the oldest university, dating back to 1636. It is a sobering experience, I continued, for me to realize that over three hundred years before the founding of Harvard, the University of Padua was already a functioning university, rich in its scholarship in the arts and the sciences. Harvard is but one link in a chain of intellectual endeavor, the heir of a continuing pursuit of universal understanding.

Gobbo's interest in my writings persisted after my lectures were over and my wife and I had returned home. In one of my published essays, I had used the phrase "fiddler on the roof" as a metaphor for the fragility of human life and the courage required to be creative in the face of such fragility. She wrote to me from Italy shortly after my return to the States, to ask for an elaboration of my meaning in using this phrase, and to inquire where the phrase had originated, suggesting that it might well have come from the writings of the Jewish author Sholom Aleichem, with a later assist from Chagall, whose paintings do at times depict a fiddler floating above a rooftop in the shtetl.

I inclined to agree with this suggestion but she wanted to nail it down. Consulting two authorities on linguistics and Yiddish literature in the United States, I was told that the phrase definitely does not appear in the writings of Sholom Aleichem, although the Chagall influence may well have been in the author's mind, whoever he was. This negative result led to further detective work by Gobbo, who got the libretto of "The Fiddler on the Roof" from the library to see if there was a clue there. Unfortunately, there wasn't, so I decided to write to Sheldon Harnick, the librettist of the musical, to ask how the title originated. In return, he replied that he no longer remembered who came up with the title, which emerged after discussion with his colleagues. They had wanted a title suggestive of Chagall and the shtetl background where the musical is set, suggestive also of the fragility of shtetl existence for Jews, but at the same time expressing the striving for joy, symbolized by the fiddler's music.

Gobbo, who had, in process, read a number of my essays and found interesting my notion of conversation as a model of teaching, proceeded to write a book on my work, entitled *La Conversazione come Metafora dell' Educazione*.[6] A delightful benefit of this work for me was the opportunity it gave me of extending my rudimentary knowledge of Italian by reading her text with the references to my own writings providing a constant anchor for my interpretation. My friendship with her has continued not only through correspondence but also through her visits to Harvard, one summer when she attended a Summer institute of mine and, for an extended time, as a Visiting Scholar at my Philosophy of Education Research Center, where we had the opportunity for frequent conversations on theoretical topics and where she presented a colloquium on her anthropological research into the education of fairground and circus children.

CHAPTER 13

EDUCATION PROJECTS

One day in 1979, I received a telephone call from Max Rowe, who, as I later learned, was a chief executive of the British Rothschild Foundation based in London. He asked to see me and I suggested we meet at my home the following day. It turned out he had a proposal to make to me. The Foundation, which had a long-standing interest in social and educational development projects in Israel, had been approached by the Israeli Ministry of Education with a problem for which they sought the Foundation's help.

With the sponsorship of the Ministry, Israel had just authorized an extra year of compulsory education, with a longer school day, to be added to its school system at the junior high level. They requested the Foundation's help in devising an appropriate curriculum for that extra year which, they hoped, would integrate informal and formal schooling. The Foundation had been favorably disposed toward the ministry's request, and was setting about to find American educators willing to help with this task. Mr. Rowe had first consulted with some of the faculty at Teachers College Columbia. They had shown no interest in the project, but had referred him to me. He thus asked if I would be willing to study the feasibility of integrating formal with informal education and, in a second phase, to propose pilot projects to test ideas emerging from the study.

I was disinclined to accept such an assignment and told this to Mr. Rowe. He asked me, however, not to make a final decision but to wait until June when I could discuss the matter in person with Sir Isaiah Berlin who would be visiting Boston to receive an honorary degree at Harvard's commencement. I agreed to hold off any final decision until after my meeting with Sir Isaiah, which I anticipated with mixed feelings. In part, I looked forward to it, since I respected him and welcomed the opportunity to get his opinions of the Israeli educational and social situation. In part, as well, I dreaded the meeting, knowing of Berlin's legendary powers of persuasion and recognizing that it would be hard to give him a negative reply.

Commencement day that June turned out to be a scorcher, sunny, unusually hot and humid. I had invited Berlin to meet me after the morning festivities in my ground-floor Emerson Hall study. My study had no air-conditioner but I had my two windows and my door wide open and had removed my jacket and tie. I was grateful that I had had the foresight to put on a short-sleeved shirt that morning. Thus I felt able to tolerate the heat although I was certainly uncomfortable.

Sir Isaiah arrived precisely at the appointed time. We had met briefly at least once before during a visit of his to Harvard and had corresponded at least once concerning a prospective fellow of Wolfson College, Oxford, of which he was the director. I had of course read a number of his writings and heard two brilliant lectures of his, one on freedom, the other on Vico, on both occasions of which he held his audiences spellbound with his erudition as they struggled to keep pace with his rapid-fire and mumbled delivery. I had also known of his positive Jewish identification, his liberal Zionist sympathies, and his advisory role to the Rothschild Foundation's projects on behalf of Israel. Thus, anticipating this first in-depth discussion with a thinker I much admired, I welcomed him to my sweltering study.

He was dressed in formal English style, dark wool suit and vest. Knowing that he had spent the morning clad in Commencement robe as well, I urged him to remove his jacket and vest. But he firmly refused, insisting on preserving his formal appearance during our hour-long talk. I told him about my reluctance to take on the project that had been outlined to me by Mr. Rowe. In part, I felt that practical curriculum design was beyond my competence; in part I was uncomfortable with the political situation in Israel.

On the latter point, he made clear that he shared my misgivings but argued that the project in question was likely to help develop a wide range of talents in the youth, open opportunities to the heretofore disadvantaged, and improve and liberalize the education available to all. I felt the force of his points but told him I was not able to decide at the moment, promising that I would ponder the idea and respond in the near future by letter. He seemed satisfied that I had not rejected the proposal outright and that I intended to give the matter further thought. With that, he said goodbye and departed into the sultry outdoors, still wearing his heavy garb.

After thinking the matter over, I wrote to Mr. Rowe to tell him of my conclusion. I had decided, in short, not to accept his proposal. First of all, I had had no practical experience in curriculum development. Secondly, even if what was wanted was only a recommendation of guiding principles for the Israeli schools, I felt that it was inappropriate for anyone from another country to offer recommendations to the educators of Israel as to how to manage their own system. I had known of cases where American so-called experts, invited to other countries to offer advice, were only too glad to pronounce on the proper solutions to these countries' problems, making up in arrogance what they lacked in understanding of the context.

However, I offered a counter-proposal for the Foundation's consideration. I was due for a sabbatical in the coming term which I thought I might usefully spend in exploring the local educational scene in order to identify school practices which struck me as novel and promising. I could write up reports on these practices and then come to Israel where I would describe them to an audience of Israeli educators and hear their descriptions of their own problems and projected solutions. Following such presentations and reflecting on our ensuing discussions, I would be prepared, upon my return home, to write a final report to the Rothschild Foundation with recommendations of particular curriculum initiatives that seemed to me worth consideration. It could then decide, in consultation with the Israeli educators, which, if any, of the recommendations to pursue.

Mr. Rowe, after consultation with the Rothschild board and particularly with Sir Isaiah, wrote to say that the Foundation had accepted my proposal. Thus, I set about making plans to actualize my idea. With Dr. Kenneth Hawes, a doctoral advisee of mine whom I hired as a research assistant, I made a tentative list of local educational practices or projects that seemed to us worthwhile exploring, and we set up a schedule of visits which we would undertake to the venues of such practices or projects.

The following several months were spent in making such visits and preparing preliminary reports of each visit. These months turned out to be unusually instructive and gratifying in bringing us into contact with a number of quite ingenious educational practices and the unusually intelligent educators who had developed them or carried them out. The philosophy of education, which I had been teaching in the abstract, was here brought to earth in the practicalities of school practice and contributed greatly to my further education.

Hawes and I prepared ten case reports of local practices we thought worthwhile in themselves and of potential relevance to the concerns of Israeli educators as articulated by the Rothschild Foundation. To illustrate the sort of interest we found in the practices we investigated, let me describe one of the sites we visited, namely, The Cambridge School at Weston, an independent boarding and day school in a country town west of Boston. The description that follows is of the school at the time we visited it, in 1980.

The initial reason for visiting the school was the important place given the arts in the life of the school, a fact made strikingly evident by the position of the arts studio at the center of the school buildings clustered around it on the periphery. The key structural feature that enabled the unusual arts emphasis was the module system dividing the year into seven mini-terms of four and a half weeks each, replacing the conventional division into two semesters of four and a half months a piece. The school day consisted, moreover, of three blocks of one and a half hours each, a longer period than the conventional one of fifty minutes each. Normally, a student was expected to take, in each module, one course continuing through two modules and one course taking up one module only, although variations were possible. Such flexibility provided longer class sessions for immersion in a subject, particularly helpful in working at studio arts. It also encouraged cross-disciplinary collaboration by faculty members and their experimentation with team-taught modules. Finally, such flexibility attracted practicing artists to the faculty by allowing them to concentrate their modules in a segment of the school year, leaving the rest free for their creative art work. The combination of short modules and lengthened class periods tended also to keep student interests alive throughout the year, giving greater opportunity for absorption and intensity of involvement in each class.

The school was not a specialized art high school, but rather a school with a regular rounded curriculum, which thus favored integration of the arts with other studies. All the arts teachers were practicing artists but a most unusual feature was the fact that the school had a resident dance company whose members functioned also as teachers of dance. The school had in fact built a special studio to accommodate the company as well as to facilitate the instruction.

Hawes and I visited a dance class taught by Martha Gray, who had been affiliated with the Dance Center at Harvard and was head of the company. We also had an extended discussion with her following the class. She struck us as clearly a brilliant teacher as well as a dedicated dancer. Her teaching exemplified high standards of performance, seriousness of purpose, subtlety of perception and sensitive interaction with her students. We felt that no one who had had the benefit of such examples could easily adopt the vulgar, sentimental, or demeaning conceptions of the arts that, unfortunately, still have currency in various quarters in education.

The educational ingenuity evidenced by the faculty under the system we have described is illustrated by Gray's discussion of collaborative teaching at the school. She had emphasized the flexibility of the module arrangement in encouraging pairs of faculty members to teach courses as teams. Somewhat dubious about the extent to which such collaborative teaching was feasible, I remarked that I could envisage such collaboration pairing the subjects of history and English, or physics and mathematics, or chemistry and biology, but could there be a pairing between electronics and dance, for example?

To my surprise, she immediately replied that there had, in fact, been such a pairing in a recent school year. It seems that the physics students who had been studying electronics rigged up a set of beams and motion sensors around the perimeter of the dance studio, each sensor responsive to a beam emitted across the dance floor and equipped to sound a tone of a particular pitch when its beam was broken. The dance students, in their turn, choreographed a system of movements which, involving the sequential breaking of specified beams, produced a complex musical pattern as they executed these movements in dance. One can easily imagine the creative difficulty posed to students by this pairing as well as the inventiveness and discipline required to carry through the project involved.

Having completed the ten reports of local practices I had considered promising, I sent these reports on to the Rothschild Foundation and they were also made available to several Israeli educators in advance of my visit. In Israel, I met with about thirty such educators, both individually and in seminar meetings, to discuss the practices I had described in my reports and to hear accounts of the Israeli projects in process or in the planning stage. In addition, I visited, aside from university settings, also the Rubin Academy of Music, the Tel Aviv Center for Educational Technology, the Bezalel Institute of Fine Arts, and several other educational institutions.

One of these institutions was the Haifa Municipal Theatre, under the artistic directorship of Mr. Oded Kotler, to whom I had been referred by Dr. Nola Chilton of the Tel Aviv Drama department. She had been visiting at Brandeis University during the year prior to my departure for Israel, so I was able to meet her in Boston before my trip. A most unusual creative artist, Chilton had become well known for her pioneering development of documentary theatre, which had radically altered the state of Israeli drama. Together with Kotler, she had stimulated the growth of an indigenous theater responsive to Israeli realities and capable of communicating with the young and with minority and disadvantaged communities.

Chilton's method was to interview members of these communities, recording their words on tape. Supplementing these tapes with letters and diaries, she employed her subjects' own expressions in her scripts, treating her actors as voicing instruments for those who themselves had no voice. The role of theatre, for Chilton, was not to create illusion on stage but to bring reality to the audience, provoking recognition of problems and stirring thinking and conscience. Her aim was to replace vulgarization, sentimentality, and stereotyping with the dignity of authentic expression.

The work that especially impressed me was her experience in Kiryat Shemonah with the mainly Oriental Jewish population, that is, those who had immigrated to Israel from the Arab countries. Here she and Kotler had brought a troupe of ten actors who went to live there for an extended period, bringing theatre into the buildings and community houses of the local community and sharing their daily lives. When we arrived in Israel, we arranged to meet Kotler in Haifa along with groups of young actors who were living in the Neve Yosef neighborhood of Haifa and working with the mainly Oriental community there.

What was most striking about the work of these actors was that they had extended their dramatic role in a radical way, becoming teachers and social workers as well. Before each performance, which they typically staged in the back yards of the buildings where the families lived, they knocked on the townspeople's doors, personally inviting them to attend, since most had never seen a stage performance before and hardly knew what to expect.

In time, they invited the teenagers, including those who were difficult or troubled, to take part. Beyond dramatic performances, the actors also worked in primary and secondary schools with children, held improvisation sessions for disaffected youths, met with adolescent clubs and mothers' groups for discussion of their problems and developed creative drama classes. By all accounts, they achieved an extraordinary rapport with this community.

What these actors, under Chilton's and Kotler's leadership, had done was to invent a hybrid role of actor-teacher and social worker, opening up new educational and career options for some students and potentially encouraging new forms of teaching in the schools. Stimulating communication with disaffected or rebellious youth seemed, further, likely to promote improvement in their prospects of constructive self-development. This contact with the Chilton and Kotler work, a byproduct of my Rothschild project, was for me a genuinely instructive, indeed inspiring, experience.

I have mentioned the benefits of interacting with academic colleagues outside of one's own university. Another form of interaction, especially important for philosophy, is the interaction with life outside the academy. As an admirer of Dewey, who valued concern with the problems of men above concern with the problems of philosophy, I felt the pull of such interaction. Trained as an analytic philosopher, I had for a long time deplored the increasingly narrow focus of analytic professional journals, while acknowledging that specialization is prerequisite to scholarly progress. My teaching in an Education school as well as a department of philosophy enabled me to try to lead a double life, not by pursuing analysis separately from my concerns with education, but by striving to apply analysis to such concerns in addition to tackling analytic problems in their own right.

Still, the effort to conduct such a double life was always challenging, and new opportunities to bridge its dual goals ever welcome. Thus, when the National Endowment for the Humanities announced a program, in the early seventies, that promised a new way of applying philosophical analysis to education, I became interested at once. The point of the program was to strengthen the role of humanistic studies in the continuing training of mid-career educators. The Endowment invited university representatives of the humanities to submit proposals for short-term summer institutes for which mid-career educators would be eligible to apply and, once chosen, would be given a suitable stipend for the occasion.

The idea appealed to me upon reflection as offering a challenging opportunity for me to apply my analytic background to the education of mature professionals, supplementing my normal teaching of graduate students about to embark on their careers. Accordingly, I submitted a proposal to the Endowment to conduct a two-week summer institute at Harvard in the field of philosophy, centering on the theme, "Freedom and autonomy in recent educational thought". My proposal was accepted, and I set about trying to organize a promising curriculum for the fifteen or so participants I would choose from among the expected applicants which, in the event, numbered well over sixty.

The plan I finally decided on was to take as the main text for the Institute Kant's *Foundations of the Metaphysics of Morals*.[1] I proposed to introduce the topic and the text in the first hour or two and thereafter to proceed as follows: Seminar meetings would be held for two hours every weekday morning, with afternoons free. For every session after the initial couple of orientation lectures, each participant was to have read an assigned section of the Kant text and written a paragraph or two on the section in question. The session would be spent in having several of these paragraphs read aloud to begin, with general discussion taking over thereafter. The prepared paragraphs were totally open; they could consist in criticism, argument, application, generalization, etc., as long as they took off from a close study of the section in question.

Now, the Kant text is a short one, but it is also quite difficult and I recruited my participants in such a way as to represent a wide diversity of educational careers, both in teaching and in various levels of administration but in no case requiring any background in philosophy, my purpose being to connect a tough analytic discourse with extensive practical experience in education.

Colleagues to whom I revealed my plan in advance were, without exception, skeptical; they in fact thought my curriculum would be a disaster. First of all, they thought the text would be impenetrable to people without prior philosophical training. Secondly, they were convinced that the participants would be unwilling to put in the work required daily, of trying to comprehend and to write about difficult and unfamiliar material. Thirdly, they argued that, coming to a summer institute immediately following a heavy academic year of teaching or administration, participants would not be able to muster the energy required for the institute's difficult demands. These various considerations gave me pause but failed to deter me, determined as I was to follow my plan and take the consequences.

My determination was not mere stubbornness. It was based on a faith, born of many years of teaching experience, that philosophy, properly presented to adults, would ignite their interest and motivate their serious thought. I had, it is true, been used to teaching university students, whose primary occupation was study. Here, I was going to be teaching professional practitioners, likely to be impatient with ivory tower speculation, and drained of psychic verve by this preceding year of hard labor in the vineyards of education. Thus, my faith carried less than utter conviction; my plan was risky and though I began with confidence, I also had qualms.

As it happened, the institute exceeded my wildest expectations. The participants who arrived from all corners of the United States came with an adventurous spirit but also with some trepidation. Nevertheless, after my introductory pair of lectures, they began increasingly to interact with Kant's text in earnest, discussing his ideas, arguing with him, relating the issues to their own experiences and so forth. Despite their lack of prior philosophical training, they found Kant's text not only understandable but intriguing, of existential as well as theoretic import.

Far from lacking the energy to engage theoretical problems after the heavy practical work of the preceding year, they reported finding the institute's theoretical emphasis energizing, in providing them with an opportunity to step back from the urgencies of their professional practice and to think through the fundamental assumptions of such practice in the company of their colleagues. They spent long hours in the afternoons, writing their paragraphs for next morning sessions and discussing their opinions with other participants. And, since their views of the passage to be discussed had been crystallized in the paragraphs already prepared for the day, seminar discussions were never dull, but proceeded through lively argumentation by the participants, with only moderate intervention by myself.

The institute proceeded in this manner throughout, completing Kant's text and interweaving some related recent papers on occasion. There was one difference of assignment for the last session: The participants were asked to review all the paragraphs they had written earlier, to edit them for consistency and revise them so as to accord with their latest considered opinions. Thus each participant would be able to leave with an essay representing his or her philosophy of education bearing on issues of freedom and autonomy.

The Endowment was so pleased with the Institute as assessed through a site visit and post-institute evaluations by the participants and myself, that they were willing to entertain a repeat in the following year, and again in a third year. Thus, I in fact gave two more institutes sponsored by the Endowment in the two summers succeeding the first one in 1976 and hoped, in the series of three, to have demonstrated the power of philosophy to speak to the professional practitioner and to elicit a comprehending response.

This connection of philosophy with the professions has been a long-term preoccupation of mine as a result of my university position in Education and philosophy, and I have taken whatever opportunity I could to promote such connection both in writing and in person. One such additional opportunity happened my way as a result of the work of a post-doctoral advisee of mine. This advisee, Dr. Uwe Steucher, was a mid-western psychology professor who had earlier written a book on autism and was especially interested in working with children in the oncology ward of Children's Hospital. There, my colleagues and I managed to get him an introduction to the medical staff, whose members encouraged him to work with the patients in his own innovative way.

He spent long hours, day and night, in the ward, with children who had cancer and had not long to live. He tried to emphasize the agency of the children, limited as it was, in whatever way he could. He brought his guitar to the ward and taught them songs; he engaged in conversation with them about longevity, emphasizing its relative nature and his own mortality; he brought them small souvenir blocks of California redwood trees, to illustrate their longer life compared to that of human beings. He also spent many hours in the company of grieving parents who, he said, invariably were much more devastated than their children by the prospect of their children's death. All in all, he did unusually impressive work at the hospital, which appreciated it enormously and wanted him to stay on after his Harvard year was completed–a request he could not fulfill because of prior obligations.

One of the innovations Steucher introduced at the hospital during his year there was an early morning seminar for the medical staff dealing with human issues relating to their work. As he explained it to me, he believed that such a seminar was especially important for medical professionals working closely and constantly with mortal illness and bereavement. Having to control their own natural responses to the illnesses, pain, and dying of their patients and the grief of parents, the physicians and nurses tended to defend against emotional reactions by transforming human issues into technological ones, thus providing the distance required for them to be effective. But Steucher thought that such defense was dangerously stressful and that the human emotions ought to have some form of release. This was the point of his new seminar for hospital staff, and I understood its rationale and agreed with it.

However, I was somewhat taken aback by his urgent request to me to conduct one of the forthcoming seminars as a philosopher, on the topic of death. I had never thought much professionally about the philosophy of death, concerning which there is in fact a substantial literature. But I admired Uwe and his work so much that I found it impossible to say no to him. Thus it happened that on an icy and snowy morning, I drove myself to the hospital at an early hour before medical rounds, to conduct a seminar for a roomful of physicians, technologists and nurses on a topic I labeled "The Question of Death".

I began hesitatingly but grew in confidence as I realized the intense and alert interest of my hearers in what I was saying. Inviting their discussion, I attempted to draw them into a consideration of the many human issues surrounding death and dying. The response was overwhelming. It was as if a dam had been breached. The discussion grew more and more animated, more emotional in tone and increasingly self-revelatory as the seminar went on. When our limited time was up, the proceeding had perforce to end, but the participants were not finished. Most had to leave the discussion in mid-air, but several followed me down the corridor and out to the exit, continuing to carry on the interchange in which they were so involved. The whole experience remains in my memory as an indication of the power of philosophy conceived as a discussion, with professionals, of the human issues raised by their practice.

CHAPTER 14

PHILOSOPHY OF EDUCATION RESEARCH CENTER

I have already indicated how my academic appointment in Education as well as in Arts and Sciences reinforced a broad conception of my work. Unlike colleagues in departments of philosophy, who continue to teach and advise primarily undergraduate concentrators and graduate students in the subject, I taught and advised both such students and also those preparing to teach the various school subjects or aspiring to administrative or policy positions within educational systems. These tasks brought me into close touch with the whole range of the school curriculum as well as the problems and preoccupations of educational professionals.

My interest in the work of professionals was illustrated in the last chapter by the institutes I conducted for the National Endowment of the Humanities. The involvement with school curriculum may be indicated by a paper I wrote in 1970, entitled "Philosophy and the Curriculum", which argued that the training of any teacher of a school subject ought to include a serious acquaintance with the philosophy of that subject.[1] In the graduate lecture course I taught at the School of Education, I regularly distributed a listing of selected books in the philosophy of each teaching area, e.g. science, mathematics, history, literature, art, music, etc., and required each student to write a final paper reflecting both the student's experience in teaching his or her subject and evidence of having thought through some philosophical question pertaining to it. The interpretation of philosophy's role here as "philosophy of—" thus conceived it to be as broad as any area of the arts and sciences, devoted to deepening its understanding and promoting appreciation of its import.

Analytic philosophy has been regularly attacked for its insularity. I have myself criticized the overly restricted range of much of what is published in analytical journals. But analysis is, in any case, not a school of philosophy but rather a pursuit of clarification and understanding, using the most effective logical instruments of thought available. That much recent analytic publication has been too restricted is a shortcoming not of analysis but of analysts. Analysis itself, applicable to every area of enterprise, is limited only by human interest and ingenuity.

Trained in analysis, I was fortunate to have been educated, early and broadly, not only in general subjects but also in Jewish culture, its classical origins, sacred texts and languages, interpretive literature, and varieties of religious thought. In the process of my early study in the family, in the yeshivot and later at the Jewish Theological Seminary of America and at Harvard, I moreover came into contact with diverse scholars of prodigious talent and achievement in various areas of Jewish study which shaped my sensibility independently of my professional specialization in contemporary analysis.

There is no doubt that my Jewish education predisposed me to inclusive humanistic attitudes in the very practice of analytic interpretation.

In my graduate study of philosophy, I was again fortunate to have studied with philosophers who applied analytic modes of thought over a broad range of topics, from mathematics and natural and social sciences to history, politics, literature and the arts, aside from traditional philosophical problems. Such breadth of scope, associated with pragmatism, characterized not only my early graduate teachers Ernest Nagel and Sidney Hook, who were influenced by Dewey, but also the teacher who most deeply influenced my graduate career, Nelson Goodman. Goodman's work, superbly analytic, elegant and nuanced, ranged over fundamental issues in applied logic and the philosophy of science and language, in the theory of knowledge, in metaphysics, and in the theory of art and literature. In promoting my candidacy for a teaching position at the Harvard Graduate School of Education, he was well aware of the low esteem of education studies at the university, but was staunch in his conviction that the pursuit of philosophical understanding is applicable in every domain and certainly called for in education. This conviction was enormously encouraging to me as I started upon my philosophical career and has remained a principle of mine throughout.

In addition to teaching both students of philosophy and students of education, I sought to make contact with established scholars. Charged by Dean Theodore Sizer with chairing a committee to draft a development report for the Harvard Graduate School of Education, (later published in 1966 as *The Graduate Study of Education*),[2] I included a proposal to establish a Whitehead Fellowship program at Harvard that would welcome post-doctoral educators of diverse interests for a year's residence devoted to independent research. This program was established by faculty vote on the report and flourished for several years, until the grant we received ran out of money and failed to attract further funding.

My collaboration with Lawrence Cremin in the founding of the National Academy of Education, already described, was inspired by Robert Ulich's vision of an association of mature scholars elected by merit from all relevant disciplines bearing on education. The Academy, duly established in 1965, continues to function, with annual meetings as well as a program of committee reports on problems of educational research and policy.

Two special purpose projects at Harvard in which I participated also provided an environment in which to pursue interdisciplinary research in education. One was Project Zero, founded in 1968 by Nelson Goodman, the purpose of which was to investigate basic abilities required in the arts. Adopting this organizing purpose, Project Zero centered on ideas growing out of Goodman's book, *Languages of Art*,[3] which had been published not long before. The other project, the Project on Human Potential, was initiated in 1979 by a request from the Bernard van Leer Foundation of the Hague to the Harvard School of Education to investigate the state of our knowledge concerning human potential and its realization. This project, conducted by a group of Harvard scholars over a number of years, explored the concept of potential and its applications from a philosophical, psychological, anthropological, and cross-cultural point of view.[4]

These various associations served as background models for me when, in 1983, along with my colleague at the time, V.A. Howard, I decided to see if we could develop a post-doctoral research center in Education at Harvard. The motivation was to establish an ongoing presence of experienced scholars from several countries, with a broadly philosophical point of view, approaching education from a variety of perspectives. Unlike the Whitehead Fellowship program, it would emphasize the philosophy of education in particular. Unlike the National Academy, it would be centered at the Harvard School of Education and provide a continuing community there not limited to annual meetings. And unlike Project Zero and the Project on Human Potential, it would not have a special research theme, but would be open to any topic bearing on education and amenable to philosophical consideration.

The intent to include visiting scholars from other countries arose from my feeling that educational issues in the United States, and at Education schools specifically, were being treated from an overly local and present-centered point of view, focused mainly on urgent problems of contemporary school practice. Extending the cross-national and historical range of studies to be welcomed at our projected center would, we thought, help to overcome the provincialism of Education schools' approaches to education and thereby deepen and sophisticate the grasp of the subject available to students, faculty and scholars alike. In general, we sought to cultivate philosophical treatments of the subject, emphasizing conceptual, theoretical, cultural and moral issues, without restriction to any philosophical school or method. We hoped thereby to encourage the breaking down of encrusted educational divisions such as those separating science and the humanities, pure and applied sciences, liberal and technical education, cognitive and affective education, and formal and informal education, by emphasizing underlying mental processes rather than current school practices and prevalent curricula.

Armed with this ambitious set of ideas, Howard and I still needed a name for the prospective entity we were hoping to bring into being. Eventually, we chose to call the yet unborn baby "Philosophy of Education Research Center", taking advantage of the jaunty acronym "PERC". The practical problem now to be faced was how to bring it to birth, a process for which merely philosophical midwifery had left us totally unprepared.

Proceeding from total ignorance, we decided to forage for facts. Since the age of the internet had not yet fully dawned, we used the traditional method of working from books, scanning several directories of foundations to determine which foundations might have an interest in philosophical approaches to education. Having found about thirty of these "possibles", we sent the same letter of inquiry to each, giving a brief account of our idea and asking if the foundation in question might be willing to support such an endeavor, in which case we would be glad to send a more detailed outline of our plan. A few replies came back soon, all negative, while no other replies came back at all, with one exception that reached us after two or three months. This reply, from the Exxon Education Foundation, was positive and moreover enclosed a check to provide us support for three years, in advance of any further clarification from us.

The delay in this reply's reaching us was the result of our total naiveté in matters of university fundraising. We had not known that applications for funds by university branches had to be approved in advance by a central committee, whose job it was to ensure that the same donors were not solicited by more than one such branch and that wealthy donors were reserved for grant requests of suitably large amounts. In our innocence, we had violated this norm by failing to get initial approval for our relatively small grant request, approval which we likely would not have gotten had we asked. As a result, various central administrators were miffed at us for our end run, and when the unexpected Foundation check arrived at their offices where the Foundation had sent it, it apparently took them quite a while to overcome their mixed feelings and to send it on to us. Thus, Perc was born as a result of our academic inexperience, not with a bang but with a whimper.

This inauspicious beginning in 1983 was followed by later generous support from other agencies, among them, notably, the Latsis Foundation and, for its final several years, the Mandel Foundation. Additional support for the individual researches of some of our scholars came from the John Dewey Foundation and the Spencer Foundation. The funding Perc received from such sources did not, alas, enable it to offer stipends to its visiting scholars or even to provide them with secretarial assistance or exclusive office space. Such support, however, enabled Perc itself to have offices, a secretary and a research associate. Visiting scholars came with their own funds, either grants, sabbatical salaries, or savings, and in return received university identity cards, access to the libraries, and course auditing privileges with the approval of the relevant instructors. Most important, they were invited into the Perc community to pursue their independent studies and to present public colloquia based on such studies.

Under the guidance of co-Directors V.A. Howard and myself until 1998, and of myself as Director thereafter, Perc flourished beyond our fondest expectations. Without advertising or recruitment, we were able in the twenty years of its existence to host more than ninety visiting scholars and associates from over nineteen different countries, and to sponsor well over one hundred and fifty-eight public colloquia. The topics our scholars dealt with ranged widely, touching on the sciences, the arts and the humanities, on the philosophy, psychology, sociology, anthropology and history of education as well as comparative studies of teaching and learning in different countries. In Perc's latter years, we developed an email network including current and former visiting scholars to keep them all informed of new Perc publications as well as announcements of continuing colloquia.

For four summers, from 1986 to 1989, Howard and I also extended Perc's reach to include short-term institutes designed for educators at all levels eager to step back from the pressures of daily tasks and spend time reflecting on the purposes and processes of learning. We focused these new "Institutes on Thinking: Critical and Creative" on such concepts as thinking, reasoning, problem-solving, intuition, imagination, thinking in performance, and so forth.

In a fifth summer, 1990, we conducted an institute we named "Seminar on Work and Education", devoted to deliberations on education and training, occupation, job, and vocation, liberal and technical education, art, craft, and performance. Participants in all five institutes were experienced educators who came from different parts of the country, having worked in a variety of educational capacities, all sharing a thirst for reflection on the fundamentals of their craft.

For me, Perc was a boon, especially after my retirement from active teaching in 1992, in providing me with a stimulating international environment of scholars of education, all engaged in diverse original studies. Rather than portraying certain of these scholars in depth, I will here briefly sketch several of their projects in order to suggest the combined scope of their interests.

From the beginning, Perc tried to build a historical and comparative perspective into its work, and to focus on fundamental rather than transient problems. Our Asian scholars, early on, helped us to further both aims. Minoru Murai of Tokyo presented Perc colloquia on human nature as viewed in classical Chinese thought: Is man essentially good, essentially evil, or essentially neither? Ynhui Park of Korea presented a number of colloquia comparing Asian and Western thought, dealing with metaphysical, moral and aesthetic themes, and touching also on problems of globalization. Kuniko Miyanaga of Tokyo presented colloquia addressing issues of creativity and identity with special reference to Japanese culture, and analyzing problems of globalization.

Aside from comparative studies, many Perc colloquia dealt with major aspects of learning. David Perkins of the Harvard School of Education offered accounts of the psychology of problem solving. Xu Di of the University of West Florida discussed the nature of intuition and its educational ramifications. Stephen Brown of SUNY, Buffalo, offered a colloquium on humanistic aspects of mathematics and Brian Alters of McGill presented a colloquium on biology teaching with special reference to the issues raised by creationism.

A number of colloquia were concerned with literature and the arts. Thomas Winner, emeritus professor of Slavic at Brown, discussed the literary approaches of the Prague school of linguistics and poetics, and Richard Sterne, emeritus professor of English at Simmons, presented accounts of Indian novels, both pre-colonial and post-colonial. Heidi Westerlund of Finland discussed aspects of music education, and Rifat Chadirji, an architect from Iraq, based in London, presented his views on the philosophy of architecture, while Daisy Igel, an architect from Brazil and the United States, presented a colloquium on the history and philosophy of the Bauhaus School. V.A. Howard offered colloquia concerned with concepts of art and utility in education, as well as discussions of intelligence and skill in the practice of artists.

This brief sampling of colloquium topics addressed by Perc scholars gives a sense of the collective range of their intellectual interests and accomplishments. The decision by the university administration, in June of 2003, to close Perc down, as a result of changes in the guidelines for Harvard centers, brought forth scores of unhappy responses from those who had over the years profited greatly from its colloquia, its publications, and the active presence of its scholars. Though Perc is now gone, its twenty year career is one of which its associates may be proud for it shows how a philosophical initiative may further the pursuit of educational understanding, a crucial ingredient of any humane society.

ENDNOTES

Chapter 1

1. Israel Scheffler, *Teachers of My Youth: An American Jewish Experience* (Boston: Kluwer Academic Publishers, 1995), esp. 7–8.
2. Houston Peterson, ed. *Great Teachers, Portrayed by Those Who Studied Under Them* (New Brunswick, N.J.: Rutgers University Press, 1946).
3. Joseph Epstein, ed. *Masters: Portraits of Great Teachers* (New York: Basic Books, 1981).
4. *Teachers of My Youth*, Op. Cit., p. 8.

Chapter 2

1. Bertrand Russell, *The Scientific Outlook* (New York: W.W. Norton & Company, Inc., 1931).
2. A.S. Eddington, *The Nature of the Physical World*, (New York: The Macmillan Company, 1928).
3. Herbert Dingle, *Through Science to Philosophy* (Oxford: Clarendon Press, 1937).
4. Hans Reichenbach, *From Copernicus to Einstein* (New York: The Wisdom Library, a division of Philosophical Library, 1942).
5. L. Susan Stebbing, *Philosophy and the Physicists* (London: Methuen & Co. Ltd., 1937).
6. Alfred North Whitehead, *Science and the Modern World* (New York: The Macmillan Company, 1925).
7. Alfred Tarski, *Introduction to Logic and to the Methodology of Deductive Sciences* (New York: Oxford University Press, 1946).
8. Morris Raphael Cohen, *A Preface to Logic* (London: Routledge, 1946).
9. Arthur Edward Murphy, *The Uses of Reason* (Westport, Conn.: Greenwood Press, 1972).
10. Sidney Hook, *Reason, Social Myths and Democracy* (New York: The Humanities Press, 1950).
11. Morris Raphael Cohen and Ernest Nagel, *An Introduction to Logic and Scientific Method* (New York: Harcourt, Brace and Company, 1934).
12. John Dewey, *Logic, the Theory of Inquiry* (New York: H. Holt and Company, 1938).
13. Charles S. Peirce, *Chance, Love, and Logic; Philosophical Essays by the Late Charles S. Peirce*, ed. Morris R. Cohen (Gloucester Mass.: P. Smith, 1949).
14. John Dewey, *Experience and Nature* (Chicago: Open Court Publishing Company, 1925).
15. Morris Raphael Cohen, *Reason and Nature: An Essay on the Meaning of Scientific Method* (Glencoe, Ill.: Free Press, 1931, 1953).
16. Morris Raphael Cohen, *The Meaning of Human History* (La Salle, Ill.: Open Court Publishing Co., 1947).

17. John Dewey, *Democracy and Education: An Introduction to the Philosophy of Education* (New York: The Macmillan Company, 1916, 1965). See also Dewey, *Lectures in the Philosophy of Education*, 1899, ed. R. Archambault (New York: Random House, 1966).
18. Ernest Nagel, *The Structure of Science: Problems in the Logic of Scientific Explanation* (New York: Harcourt, Brace & World, 1961).
19. Morris Raphael Cohen and Ernest Nagel, Op. Cit.
20. Israel Scheffler, *Four Pragmatists: A Critical Introduction to Peirce, James, Mead, and Dewey* (New York: Humanities Press, 1974).
21. John Dewey, *Human Nature and Conduct: An Introduction to Social Psychology* (New York: The Modern Library, 1930).

Chapter 3

1. Nelson Goodman, "The Problem of Counterfactual Conditionals," *Journal of Philosophy* 44 (1947): 113–28.
2. Nelson Goodman, *The Structure of Appearance* (Cambridge: Harvard University Press, 1951).
3. Nelson Goodman, "A Query on Confirmation," *Journal of Philosophy* 43 (1946): 383–85.
4. Nelson Goodman, "On Infirmities of Confirmation Theory," *Philosophy and Phenomenological Research* 8 (1947), 149–51.
5. Willard Van Orman Quine, *Methods of Logic* (New York: Holt, Rinehart and Winston, 1950, 1972).
6. Nelson Goodman, *Languages of Art: An Approach to a Theory of Symbols* (Indianapolis: Bobbs-Merrill, 1968).
7. Richard Rudner and Israel Scheffler eds., *Logic and Art: Essays in Honor of Nelson Goodman* (Indianapolis: Bobb-Merrill, 1972).
8. Nelson Goodman, "Sense and Certainty," *Philosophical Review* 61 (1952): 160–7.

Chapter 4

1. Wilfrid Sellars and John Hospers, eds. *Readings in Ethical Theory* (New York: Appleton-Century-Crofts, 1952).
2. Evelyn Masi, "A Note on Lewis's Analysis of the Meaning of Historical Statements," *J. Phil.*, 46 (1949), 670–4. I replied in *J. Phil.*, 47 (1950), 158–66.
3. Report of the Harvard Committee, *The Graduate Study of Education* (Cambridge: Harvard University Press, 1966).

Chapter 5

1. Frederic Lilge, *The Abuse of Learning: The Failure of the German University* (New York: The Macmillan Company, 1948).
2. Charles Kay Ogden and I.A. Richards, *The Meaning of Meaning: A Study of the Influence of Language Upon Thought and of the Science of Symbolism* (New York: Harcourt Brace Jovanovich, 1946).

3. John Bissell Carroll, *The Study of Language: A Survey of Linguistics and Related Disciplines in America* (Cambridge: Harvard University Press, 1953).
4. Alfred North Whitehead and Bertrand Russell, *Principia Mathematica* (Cambridge, Eng.: University Press, 1910, 1925).
5. Ivor Armstrong Richards, *The Republic of Plato: A New Version Founded on Basic English* (New York: Norton, 1942).
6. Ivor Armstrong Richards, *Speculative Instruments* (London: Routledge & Paul, 1955).
7. Ivor Armstrong Richards, *The Screens and Other Poems* (New York: Harcourt, Brace and World, Inc., 1960).
8. Ivor Armstrong Richards, *Tomorrow Morning, Faustus! An Infernal Comedy* (London: Routledge & Kegan Paul, 1962).
9. "The Rhythm of Education" in Alfred North Whitehead, *The Aims of Education and Other Essays* (New York: Macmillan, 1929, 1959).

Chapter 6

1. Israel Scheffler, *The Language of Education* (Springfield, Ill.: Thomas, 1960).
2. Israel Scheffler, *Beyond the Letter: A Philosophical Inquiry into Ambiguity, Vagueness, and Metaphor in Language* (London: Routledge and Kegan Paul, 1979).
3. Israel Scheffler, *The Anatomy of Inquiry: Philosophical Studies in the Theory of Science* (New York: Knopf, 1963).
4. Karl Raimund Popper, *The Open Society and Its Enemies* (London: Routledge, 1945).
5. Karl Raimund Popper, *The Poverty of Historicism* (Boston: Beacon Press, 1957).
6. Karl Raimund Popper, "Back to the Pre-Socratics," in (London: *Proceedings of the Aristotelian Society*, 1958).
7. Op. Cit., Pt. II, section 4, esp. p. 144.
8. Richard Bevan Braithwaite, *Scientific Explanation: A Study of the Function of Theory, Probability and Law in Science* (Cambridge, Eng.: University Press, 1953).
9. Willard Van Orman Quine and Nelson Goodman, "Elimination of Extralogical Postulates," *Journal of Symbolic Logic* 5 (1940)104–109.
10. Thomas S. Kuhn, *The Structure of Scientific Revolutions* (Chicago: University of Chicago Press, 1964).
11. Israel Scheffler, *Science and Subjectivity* (Indianapolis: Bobbs-Merrill, 1967, 2nd ed. Indianapolis: Hackett, 1982).
12. Imre Lakatos and Alan Musgrave eds., *Criticism and the Growth of Knowledge* (Cambridge Eng.: University Press, 1970).
13. The lectures were later published as a book: R.S. Peters, *Authority, Responsibility, and Education* (London: Atheling, 1976).
14. Israel Scheffler, *Philosophy and Education: Modern Readings* (Boston: Allyn and Bacon, 1958).
15. Scheffler, *Language of Education*, Op. Cit.
16. Richard Stanley Peters, *Hobbes* (Harmondsworth: Penguin, 1956).
17. Richard Stanley Peters, *The Concept of Motivation* (London, Routledge & K. Paul, 1967).
18. G.S. Brett, *Brett's History of Psychology*, Richard Stanley Peters, ed. (Cambridge: M.I.T. Press, 1965).

19. Richard Stanley Peters, *Ethics and Education* (Atlanta: Scott, Foresman, 1967).
20. Israel Scheffler, *Conditions of Knowledge* (Chicago: Scott, Foresman, 1965).
21. Alfred Jules Ayer, *The Foundations of Empirical Knowledge* (London: Macmillan and Co. Ltd., 1940).

Chapter 7

1. Gordon Willard Allport, *Personality: A Psychological Interpretation* (New York: H. Holt and Company, 1937).
2. Robert Henry Pfeiffer, *Introduction to the Old Testament* (New York: Harper, 1948).
3. Israel Scheffler and Cornelius L. Golightly, "Playing the Dozens: A Note," *Journal of Abnormal and Social Psychology* 43 (1948): 104–105.
4. Alfred North Whitehead and Bertrand Russell, Introduction to the Second Edition, Vol. 1, *Principia Mathematica* (Cambridge: University Press, 1925, 1950), xiii.
5. Israel Scheffler, "The New Dualism: Psychological and Physical Terms," *Journal of Philosophy* 50 (1953): 457–66.
6. Henry David Aiken, *The Age of Ideology: The Nineteenth Century Philosophers* (Boston: Houghton Mifflin, 1957).
7. Henry David Aiken, *Reason and Conduct: New Bearings in Moral Philosophy* (New York: Knopf, 1962).
8. William Barrett and Henry David Aiken, *Philosophy in the Twentieth Century: An Anthology* (New York: Random House, 1962).
9. Henry David Aiken, *Predicament of the University* (Bloomington: Indiana University Press, 1971).
10. Clarence Irving Lewis, *Mind and the World Order: Outline of a Theory of Knowledge* (New York: C. Scribner's Sons, 1929).
11. Clarence Irving Lewis, *An Analysis of Knowledge and Valuation* (La Salle, Ill.: The Open Court Publishing Co., 1947).
12. Gilbert Ryle, *The Concept of Mind* (New York: Barnes & Noble, 1949).
13. Willard Van Orman Quine, *Word and Object* (Cambridge: Technology Press of the Massachusetts Institute of Technology, 1960).
14. Willard Van Orman Quine, *From a Logical Point of View: 9 Logico-Philosophical Essays* (Cambridge: Harvard University Press, 1953, 1980).
15. Israel Scheffler and Noam Chomsky, "What is Said to Be," (London: *Proceedings of the Aristotelian Society*, 1958), 71–82.
16. See Quine, "On What There Is", in *From a Logical Point of View*," Op. Cit., p. 4.
17. Nelson Goodman, *The Structure of Appearance* (Cambridge: Harvard University Press, 1951).
18. Nelson Goodman, *Fact, Fiction and Forecast* (Cambridge: Harvard University Press, 1955).
19. Nelson Goodman, *Languages of Art*: *An Approach to a Theory of Symbols* (Indianapolis: Bobbs-Merrill, 1968).
20. Nelson Goodman, *Ways of Worldmaking* (Indianapolis: Hackett Publishing Co., 1978).
21. Israel Scheffler, "The Wonderful Worlds of Goodman," *Synthese* 45 (1980): 201–9).
22. Nelson Goodman, *Of Mind and Other Matters* (Cambridge: Harvard University Press, 1984).

23. Israel Scheffler, *Inquiries: Philosophical Studies of Language, Science, and Learning* (Indianapolis: Hackett, 1986).
24. Nelson Goodman, "On Some Worldly Worries," published by the author at Emerson Hall, Harvard University, Sept. 1, 1988. See also *Synthese* 95 (1993): 9–12. Reprinted in P.J. McCormick, ed. *Starmaking* (Cambridge: The M.I.T. Press, 1996), pp. 165–8.
25. Israel Scheffler, "Worldmaking: Why Worry?" in P.J. McCormick, ibid, pp. 171–7.
26. P.J. McCormick, Ibid, pp. 208–13.
27. Israel Scheffler, *Symbolic Worlds* (Cambridge: Cambridge University Press, 1997), p. 209.
28. Israel Scheffler, "Some Responses to Goodman's Comments in *Starmaking,*" *Philosophia Scientiae* 2 (2) (1997): 207–11.
29. Israel Scheffler, "A Plea for Plurealism," *Transactions of the Charles S. Peirce Society* 35, no. 3 (1999): 425–36.
30. Philipp Frank, *Philosophy of Science: The Link Between Science and Philosophy* (Englewood Cliffs, N.J.: Prentice-Hall, 1957).
31. Philipp Frank, *Foundations of Physics*, (Chicago: University of Chicago Press, 1946).
32. Percy Williams Bridgman, *The Logic of Modern Physics* (New York: The Macmillan Company, 1948).
33. Robert S. Cohen and Marx W. Wartofsky, *Boston Studies in the Philosophy of Science*, vol. 2: In Honor of Philipp Frank (New York: Humanities Press, 1965).
34. James Bryant Conant, *My Several Lives* (New York: Harper & Row, 1970).
35. Thomas S. Kuhn, *The Structure of Scientific Revolutions* (Chicago: University of Chicago Press, 1964).
36. Leonard K. Nash, *The Nature of the Natural Sciences* (Boston: Little, Brown, 1963).
37. James Bryant Conant, *Modern Science and Modern Man* (New York: Columbia University Press, 1952).
38. James Bryant Conant, *On Understanding Science: An Historical Approach* (New Haven: Yale University Press, 1947).
39. James Bryant Conant, *General Education in a Free Society: Report of the Harvard Committee* (Cambridge: Harvard University Press, 1950).
40. Israel Scheffler, *Science and Subjectivity*, Op. Cit.

Chapter 8

1. Israel Scheffler, *Science and Subjectivity*, Op. Cit.
2. Michael Polanyi, *Personal Knowledge: Towards a Post-Critical Philosophy* (Chicago: University of Chicago Press, 1958).

Chapter 9

1. Joseph J. Schwab, "Eros and Education: A Discussion of One Aspect of Discussion," in Westbury, I. & Wilkof, N. (Eds.) *Science, Curriculum and Liberal Education* (Chicago: University of Chicago Press, 1978) pp. 105–32.
2. Joseph J. Schwab, "On the Corruption of Education by Psychology," *The School Review* 66 (Summer, 1958): 169–84.

3. Israel Scheffler, *Reason and Teaching* (Indianapolis: Bobbs-Merrill, 1973 and Indianapolis: Hackett Publishing Co., 1988).
4. Lawrence Arthur Cremin, *The Transformation of the School: Progressivism in American Education, 1876–1957* (New York: Knopf, 1961).
5. Lawrence Arthur Cremin, *American Education,* (New York: Harper and Row, 1970–1988).

Chapter 10

1. Harry Austryn Wolfson, *The Philosophy of the Church Fathers* (Cambridge: Harvard University Press, 1964). A general account of Wolfson's life and work is given in Leo Schwartz, *Wolfson of Harvard:Portrait of a Scholar* (Philadelphia: Jewish Publication Society of America, 1978), from which I have learned much.
2. Harry Austryn Wolfson, *The Philosophy of the Kalam* (Cambridge: Harvard University Press, 1976).
3. All included in his *Philosophical Papers*; see note 6 below.
4. John Langshaw Austin, *How to Do Things With Words* (Cambridge: Harvard University Press, 1962).
5. John Langshaw Austin, *Sense and Sensibilia* (Oxford: Clarendon Press, 1962).
6. John Langshaw Austin, *Philosophical Papers* (Oxford: Clarendon Press, 1961).

Chapter 11

1. Steven Marcus, *Representations: Essays on Literature and Society* (New York: Random House, 1975).
2. Joseph Weizenbaum, *Computer Power and Human Reason: From Judgment to Calculation* (San Francisco: W.H. Freeman, 1976).
3. Herbert Weiner, *Psychobiology and Human Disease* (New York: Elsevier, 1977).
4. Israel Scheffler, *Four Pragmatists: A Critical Introduction to Peirce, James, Mead, and Dewey* (New York: Humanities Press, 1974).
5. Alexander Mitscherlich and F. Mielke, *Doctors of Infamy: The Story of the Nazi Medical Crimes* (New York: H. Schuman, 1949).
6. Israel Scheffler, *Beyond the Letter: A Philosophical Inquiry into Ambiguity, Vagueness, and Metaphor in Language* (London: Routledge & Kegan Paul, 1979).

Chapter 12

1. Op. Cit.
2. Op. Cit.
3. Ernst Cassirer, *Language and Myth* (New York: Harper & Brothers, 1946).
4. Ernst Cassirer, *An Essay on Man: An Introduction to a Philosophy of Human Culture* (New Haven: Yale University Press, 1944).
5. Ernst Cassirer, *The Philosophy of Symbolic Forms* (New Haven: Yale University Press, 1955–57).

6. Francesca Gobbo, *La Conversazione come Metafora dell'Educazione* (Verona: Morelli, 1990).

Chapter 13

1. Immanuel Kant, *Foundations of the Metaphysics of Morals* (Indianapolis: Bobbs-Merrill, 1959). German text originally published in 1785.

Chapter 14

1. Israel Scheffler, "Philosophy and the Curriculum," in *Reason and Teaching* (Indianapolis: Bobbs-Merrill, 1973), pp. 31–41.
2. Report of the Harvard Committee, *The Graduate Study of Education* (Cambridge: Harvard University Press, 1966).
3. Nelson Goodman, *Languages of Art: An Approach to a Theory of Symbols*, Op. Cit.
4. The books that emerged from this project were: Howard Gardner, *Frames of Mind* (New York: Basic Books, 1983), Israel Scheffler, *Of Human Potential* (Boston: Routledge & Kegan Paul, 1985), Robert A. LeVine and Merry I. White, *Human Conditions* (Boston: Routledge and Kegan Paul, 1986), and Merry I. White and Susan Pollak, eds. *The Cultural Transition* (Boston: Routledge and Kegan Paul, 1986).

Photograph Gallery

Portraits of the Scholars

Painting by Edwin Burrage Child 1929
Courtesy of Columbia University Archives—Columbiana Library

John Dewey

Courtesy of Columbia University Archives—Columbiana Library

Ernest Nagel

Courtesy of New York University Archives, Photographic Collection

Sidney Hook

Courtesy of Harvard University Archives (HUP-1N)

Philipp Frank

Courtesy of Professor Catherine Z. Elgin

Nelson Goodman

Courtesy of Harvard University Archives (HUP-3N)

Willard Van Orman Quine

Courtesy of Harvard University Archives (HUP-18a-N)

Harry Austryn Wolfson

Courtesy of Harvard University Archives (HUP-4aN)

Robert Ulich

Courtesy of Harvard University Archives (HUP-6N)

Ivor A. Richards

Courtesy of Harvard University Archives (HUP-1)

Henry David Aiken

Photograph by Ansel Adams. Used with permission of the Trustees of The Ansel Adams Publishing Rights Trust.

Ralph W. Tyler

Photograph by John Brook
Courtesy of Harvard University Archives (HUP-4a)

Francis Keppel

Courtesy of Harvard University Archives (HUP-7)

John W. M. Whiting

Courtesy of Professor R.S. Peters

Richard S. Peters

NAME INDEX

Philosophy and Education

1. C.J.B. Macmillan and J.W. Garrison: *A Logical Theory of Teaching.* Erotetics and Intentionality. 1988 ISBN 90-277-2813-5
2. J. Watt: *Individualism and Educational Theory.* 1989 ISBN 0-7923-0446-2
3. W. Brezinka: *Philosophy of Educational Knowledge.* An Introduction to the Foundations of Science of Education, Philosophy of Education and Practical Pedagogics. 1992 ISBN 0-7923-1522-7
4. J.H. Chambers: *Empiricist Research on Teaching.* A Philosophical and Practical Critique of its Scientific Pretensions. 1992 ISBN 0-7923-1848-X
5. I. Scheffler: *Teachers of My Youth.* An American Jewish Experience. 1995 ISBN 0-7923-3232-6; (Pb) 0-7923-3236-9
6. P. Smeyers and J.D. Marshall (eds.): *Philosophy and Education: Accepting Wittgenstein's Challenge.* 1995 ISBN 0-7923-3715-8
7. J.D. Marshall: *Michel Foucault: Personal Autonomy and Education.* 1996 ISBN 0-7923-4016-7
8. W. van Haaften, M. Korthals and T. Wren (eds.): *Philosophy of Development.* Reconstructing the Foundations of Human Development and Education. 1997 ISBN 0-7923-4319-0
9. N. Aloni: *Enhancing Humanity.* The Philosophical Foundations of Humanistic Education. 2002 ISBN 1-4020-0961-5
10. D. Bridges: *Fiction written under Oath?* Essays in Philosophy and Educational Research. 2003 ISBN 1-4020-1083-4
11. K.R. Howe: *Closing Methodological Divides.* Toward Democratic Educational Research. 2003 ISBN 1-4020-1164-4
12. J.D. Marshall (ed.): *Poststructuralism, Philosophy, Pedagogy.* 2004 ISBN 1-4020-1894-0
13. I. Scheffler: *Gallery of Scholars.* A Philosopher's Recollections. 2004 ISBN 1-4020-2679-X

KLUWER ACADEMIC PUBLISHERS – DORDRECHT / BOSTON / LONDON